COMMON SENSE METHODS FOR FOOD PROCESSORS

Managing
Food Industry
Waste

COMMON SENSE METHODS FOR FOOD PROCESSORS

Managing
Food Industry
Waste

ROBERT R. ZALL

Blackwell
Publishing

Blackwell Publishing Professional
2121 State Avenue, Ames, Iowa 50014, USA

Orders: 1-800-862-6657
Office: 1-515-292-0140
Fax: 1-515-292-3348
Web site: www.blackwellprofessional.com

Blackwell Publishing Ltd
9600 Garsington Road, Oxford OX4 2DQ, UK
Tel.: +44 (0)1865 776868

Blackwell Publishing Asia
550 Swanston Street, Carlton, Victoria 3053, Australia
Tel.: +61 (0)3 8359 1011

Authorization to photocopy items for internal or personal use, or the internal or personal use of specific clients, is granted by Blackwell Publishing, provided that the base fee of $.10 per copy is paid directly to the Copyright Clearance Center, 222 Rosewood Drive, Danvers, MA 01923. For those organizations that have been granted a photocopy license by CCC, a separate system of payments has been arranged. The fee code for users of the Transactional Reporting Service is 0-8138-0631-3/2004 $.10.

Printed on acid-free paper in the United States of America

First edition, 2004

Library of Congress Cataloging-in-Publication Data
Zall, Robert R.
   Managing food industry waste : common sense methods for food processors /
Robert R. Zall—1st ed.
      p. cm.
   ISBN 0-8138-0631-3 (alk. paper)
   1. Food industry and trade—Waste disposal.   I. Title.

TD899.F585Z35 2004
664′.0028′6—dc22

                                                2004005966

The last digit is the print number: 9 8 7 6 5 4 3 2 1

About the Author

Dr. Robert R. Zall, born in Lowell, Massachusetts, received both a bachelor of science degree in 1949 and a master of science degree in 1950 from the University of Massachusetts, and in 1968 a Ph.D. degree from Cornell University.

He is now Professor Emeritus in the Department of Food Science at Cornell University. Prior to joining the department in 1971, he spent almost 20 years in two major dairy industry firms as General Manager for Grandview Dairies and later as Director of Research and Production for the Crowley Milk/Food Company.

At Cornell, Dr. Zall taught undergraduate courses in sanitation, food processing, and waste management. He is a member of the Graduate Faculty in two fields, Field of Food Science and Technology and the Field of Environmental Quality where he directs graduate students as chairman or committee member in advanced degree programs.

While on sabbatical leave from Cornell University in 1977-78 he worked with Express Dairies in the United Kingdom developing cottage cheese operations using UF concentrates. In 1984-85, he served as Project Director for the on-farm UF study/demonstration project carried out in California. Dairy Research Inc. funded the work in cooperation with the California Milk Advisory Board. In 1992, he was awarded an Industry Fellow in Australia and worked with the Victorian College of Agriculture, Gilbert Chandler Campus where he lectured, worked with industry, and helped initiate a master's program.

Dr. Zall has published more than 300 scientific papers and articles. He has been awarded several patents and contributed chapters to four textbooks. He is or was a member of The Society of Agricultural Engineering, American Dairy Science Association, The Institute of Food Technology, International Dairy Federation, and The International

Association of Milk and Food Sanitarians. Dr. Zall has been the recipient of awards from State Associations of Milk Sanitarians and has been given Certificates of Appreciation from the EPA.

He travels abroad extensively to work with groups and organizations carrying out food science activities. In addition to his teaching and research work, Dr. Zall serves as a consultant to private companies in the United States and abroad. He has been on the Boards of Directors of chemical and food companies and has served as a milk industry trustee on a pension and welfare fund for dairy industry employees.

Contents

Preface

This book is not an engineering text about how to build and operate wastewater treatment facilities. The narrative is a summary of some 40-odd years experience by the author in managing waste.

As an executive in two major dairy/food organizations the author draws on about 20 years hands-on experience in managing multi-plant factories. For the additional 20 years, he has been a professor of food science teaching food processing waste management to students at Cornell University. Much of his work with graduate students focused on waste management projects.

It would be nice if managers took the time to read in-depth subject matter topics of value to their companies. Unfortunately, large numbers of our managers look to literature abstracts for everything they need to know about problems. Not all issues can be summarized in a short paragraph.

The information contained in this book can be read quickly and is philosophical in nature. If the reader accepts the concepts and puts them into practice, the effort in writing the text has been worthwhile.

Introduction: Waste Control Philosophy

Be the theme fashionable or not, some managing directors of major industrial operations claim they want to create a corporate culture where there is no such thing as industrial waste. Bear in mind that the term waste means different things to different people. For those individuals who believe that everything that goes down a drain into a sewer is something that can be recycled, reused or sold, forget that the better part of idealism is realism.

Zero Discharge

The truth may well be that for food processing factories there will probably never be installations with zero discharge. Another truth should be that just because some innovative process can be created to curb or treat under-utilized food fragments does not mean management would want to carry on such an activity. History reflects the fact that at least in the United States, until EPA legislation was passed in the early 1970s with Public Law 92-500, many firms had to be almost dragged by their feet to reduce or treat waste leaving their factories. When the driving force of environmental laws dictates measures stringent enough for food plants to cease discharging wastes, management will search for economical systems to deal with waste.

What makes it interesting for those of us in the food business is that it is common knowledge that plants with mirrorlike equipment generate different amounts of waste when processing similar raw material into pretty much the same kind of finished goods. More often than we might like to admit to ourselves, management problems exist. We need only make a few site visits to some of the factories, talk with supervisors who think they manage the factories, and listen to some of their comments.

You need not be a college professor to point to the fact that we are looking at management problems.

If we query individuals about the amount of energy used per month, or ask about detergent costs to clean the operation and are told "they take care of this at the main office," we have an insight of local management authority. To get a still better overview, it is always interesting to ask about waste flows or waste strength. Not too many managers have a real feel for this end of the operation, as they might want to think this is a problem area for plant engineering. Once again we know that for the manager with whom we are speaking, waste is not looked upon as a function of plant efficiency.

"O.K.," you say, "you have my attention! Why should I want to read your book? I do manage a food-processing factory and have people on board who take care of my waste problems. We have by-product recovery and what is not recovered we send to a regulatory approved waste treatment plant. As far as I know our treatment costs are well within the values of sister plant operations."

From the author's point of view, the foregoing assumption is flawed. While there is no quarrel with the intent of such an individual who thinks like the previous manager, I believe such a person may want to rethink this kind of attitude.

Food processing plant managers, for the most part, recognize product recovery only in terms of finished food yields per ton of raw ingredients, or perhaps as in percentage of through-put amounts. They fail to regularly measure wastes as a separate and identifiable by-product. They overlook the fact that accounting for wastage on a daily basis detects material loss, some of which might be decreased. This information can be of immediate economic importance. The best way the food processing industry can treat its waste is to avoid producing it. Often improving in-plant waste abatement methods, which are less expensive and far more productive than end-of-pipe treatment, can substantially reduce a plant's waste load.

Ask the Right Questions

Much of the waste now being generated in some food processing plants can become a managed resource for producing economic credits. One approach to use is to initiate a question and answer session with plant managers using basic and simple ideas.

Ask the following questions:

1. What is being lost to the sewers?
2. How much (of whatever it is) is being lost?
3. What is the makeup or composition of the material being lost?
4. Can the loss be stopped?
5. Can the loss be reduced?
6. If you intercept and capture the lost material before it goes to a sewer, can it be used?

Very often just knowing the "what" and "how much" provides some answers and even opportunities for inserting unused liquids or food fragments into existing products. Note that from a practical viewpoint, no waste recovery system we design could ever be as effective in retrieving waste as a scheme that never allowed waste to occur.

The rationale behind asking and then answering the above questions is simple; the efforts to provide the answers evolve into the methods and tools used in quantifying both flow volumes and waste quality. Knowing the waste flow volumes and the concentration of waste parameters gives rise to a mass balance chart that accounts for total input and output. The aim of the effort is to identify the actual losses sent to drains and sewage treatment plants, to pinpoint which unit processes generate these losses, and to ferret out the hidden losses termed "materials unaccounted for". This technique, after all, involves nothing more complex than using common sense.

COMMON SENSE METHODS
FOR FOOD PROCESSORS

Managing
Food Industry
Waste

CHAPTER 1

Who Is Watching the Store?

I f I was asked to criticize food processing management practices in the United States I would point out that American managers have somewhat different constraints than our European counterparts.

It would seem on the outside that European managers are more likely to incorporate innovative processing technologies in their factories on the basis of creating better systems than to be primarily limited by ROIs (returns on investments).

Return on Investment (ROI)

While ROI is indeed a useful approach in making investments, it frequently fails to be a strategic tool rather than a tactical method for investing funds. We should point out that focusing on ROI in the case of waste reduction could get out of hand by focusing too narrowly on short-term rewards that would be beneficial to a company.

I suppose the revolution in peeling of fruit and vegetables using caustic soda versus mechanical peeling would be a case in point.

The savings in product waste alone—not counting water, heat, and waste treatment expense—go on and on making economic contributions to the canning industry long after ROI analysis ran its course.

Changes in today's economic climate, if made, need to show a profit payback in less than 2 years—and, for some investments, shorter time periods. Managers frequently complain about such chief executive officer (CEO) policy dictated by Boards of Directors.

Such logic prevents "the domino effect" in improvements from taking place in a factory situation. For example, some experience in food plants shows the full value of good waste management practices is more than just those dollars saved by lessening sewage charges. Personnel who monitor effluent in the British dairy industry, for example, report that they see a trend in better processing technology as a spin-off to waste abatement programs. Reducing wastage seems to cause a rippling effect by initiating improvement that spreads like infection in a factory when barriers of habit fall and personnel become committed to saving or reducing material wastage.

Today's managers may well have broad and in-depth experience with being able to evaluate allocation of funds made available to the factory by use of computers, but the techniques are site-specific. Computer analysis which incorporates risk, ranking of investments, dimensional analysis, and a host of software tricks of the trade to improve profits often selects the choice that seems obvious. This being the case, we ought to wonder why increasing numbers of firms are failing and even going out of business. These new tools, no matter how elegant they appear to be, still need to be questioned as being appropriate for making the best management decision to safeguard the well-being of the corporation and its employees and to be honest to consumers purchasing the product produced and sold by the firm.

Often there is a practical aspect of process engineering that falls through the cracks of investment and change that focuses on process cost to build and neglects expense to operate. Sometimes we can see where chemical engineers are prone to using established techniques of perhaps petrol-chemical plants in evaluating food-processing schemes that just do not work. Processes geared to incur the least cost relative to heat balance, shear efficiency, etc. could well wind up making products that suffer flavor, color or odor defects and perhaps even create problems with bacteriological overtones.

Of course, innovations have limits other than capital costs. It is no longer acceptable simply to build machines to perform a task without looking at and then quantifying the side effects such equipment can generate.

The real savings from the salvage exercise are probably the subtle economies produced from using some new ideas, which in turn bring about still other ideas. We can find a family of innovations now being used in the European dairy industry which is designed to save fuel, water, labor, and unused food waste that can be made into edible-grade products. Small firms, no bigger than a large dairy farm, are installing and using small space-occupancy technology like reverse osmosis and membrane plants to condense milk instead of the more conventional vacuum evaporators. This choice saves capital investment in buildings and large boilers.

People everywhere are going to have to think more and more in terms of improving "waste quality". This being so, the food industry must manage its waste. The first step is for management to find out how much waste the plant produces. A food processing plant's garbage, made up of wasted foods, spoiled packaged goods, and liquid waste, tells the plant how professional it really is. A plant's garbage is a good indicator of how well its operation is run and of which operations need attention. Just as the plant accounts for raw product received, product made, and finished goods disposition on a daily basis, it ought to account for its liquid and solid wastes, not periodically but *daily*.

It makes sense that appropriate factory personnel ought to inventory and assign reasonable unit cost values to different waste flows leaving a plant, and then plug these values into the regular cost account systems. For many, an inventory could prove to be a very profitable exercise. It is always good to know how things are running, as opposed to how they are supposed to be running. Waste inventories have a way of raising "red flags" to indicate when and where something is amiss.

Waste losses, whether processed product, cold water, hot water, food residues, or ingredients like sugar, stabilizer, or even detergents, can be measured in terms of biological oxygen demand (BOD) or chemical oxygen demand (COD) values that result from their impact as pollutants. These values can then be equated to operating expenses in terms most people understand: money.

Most food processing plants fail to measure wastes as a separate identifiable by-product. Management usually recognizes product recovery only in terms of finished food yields, either as percentages or as pounds per ton of raw ingredients. Management can exert control in the amount of water resources and waste within any given food processing plant. Supervisors must be dedicated to the task to develop positive action programs and must follow through in all cases. Management must clearly understand the relative role of engineering and management supervision in plant losses. The best modern engineering design and equipment alone cannot provide for the control of water resources and waste within a plant. Management control of water resources and waste discharges should involve *all* of the following:

1. Installation and use of a waste monitoring system to evaluate progress.
2. Use of an equipment maintenance program to minimize all product losses.
3. Use of a product and process scheduling system to optimize equipment utilization, minimize distractions of personnel, and assist in making supervision of the operation possible.
4. Use of a planned quality control program to minimize waste.
5. Improvement of processes, equipment and systems as rapidly as is economically feasible.

CHAPTER 2

Why Waste Flows Need
to Be Inventoried

Engineers and others have published a lot of information about treating food wastes. Food fragments are either soluble or not; as such, these fragments may be recoverable by-products. More often, though, they are substances that require digestion in waste treatment systems.

There is no current lack of technology for treating and disposing of food wastes as sewage if this is management's goal. The question really is what kinds of options are available to a manager's organization to deal with waste being generated in the food-processing factory.

Inventory Frequency

Technology choice depends on capital investment, operating costs, local or regional environmental restrictions, etc. These constraints are subject to both short-run and long-term concerns of the manager and the organization. It seems almost obvious to even a layperson in this area that waste ought to be inventoried at least on a daily basis, if not a "shift" basis as any other commodity in a factory. The chemical engineer who might be queried about inventory uses the term "mass balance," which is almost always used to account for the changes and locations of reacting substances. More than one operation in the food industry falls prey to the idea that phantom (unexplained) losses exist.

Daily or individual shift inventory problems can be complex in operations that have food material in different stages of processing over 24-hour periods. Some accounting practices tolerate accounting systems where managers allocate theoretical amounts of raw materials to finished goods that can be inventoried and then lump the balance to some single ingredient or process.

For example, some dairy plants processing milk into butter and skim milk powder and other products adopt accounting methods which lump total skim milk remaining after use to powder. After allocations are made to cream and butter the firm accepts powder yield results as a reflective index of loss. Loss may well have occurred during milk separation into cream and skim milk. Loss may have taken place in evaporation unit operations where skim milk is concentrated fivefold or more into condensed milk prior to drying. The truth may well be too that the dryer itself was losing powder without being monitored.

Monitoring waste is in fact a management tool. However, it seems that even though waste-producing data are being generated, the machinery to make more effective use of them at management level is not available to deal with the information. Waste monitoring can be done around the clock; supervision for it answers, Who? What? Where? and When?

Check Waste Temperature

On a simple scale I found that merely recording the temperature of the effluent could help to isolate causes of fuel waste in plants. To use this approach, temperature records have to be obtained on a number of successive days. Comparisons of the charts will begin to indicate characteristic operations. Swings in degrees can be traced back to specific operations.

Figures 2.1 and 2.2 illustrate the fluctuations in wastewater temperatures before and after recognition of the waste problem by the plant personnel. A lobe (see Figure 2.1) was noted in which the temperature in the outfall pipe rose to 150°F. The unusual temperature rise was found to be caused by the improper use of a processing vat. With simple adjustments of the steam and water valves, it was possible to completely eliminate the lobe as shown in Figure 2.2, thus reducing steam waste and water usage. In addition, a jagged erratic temperature chart was

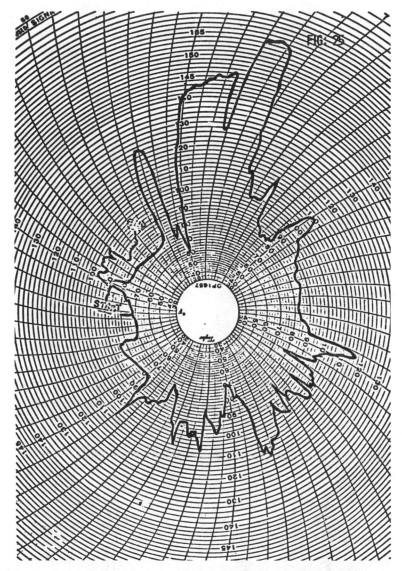

Figure 2.1. Twenty-four-hour recorder chart of effluent temperature on a typical day before conservation practices were put into effect.

formed in Figure 2.1 when waste volume was high because of careless water use. After a positive effort to reduce water use, one notes a smoothing out of peaks and valleys with creation of a more uniform temperature chart for the day's operation.

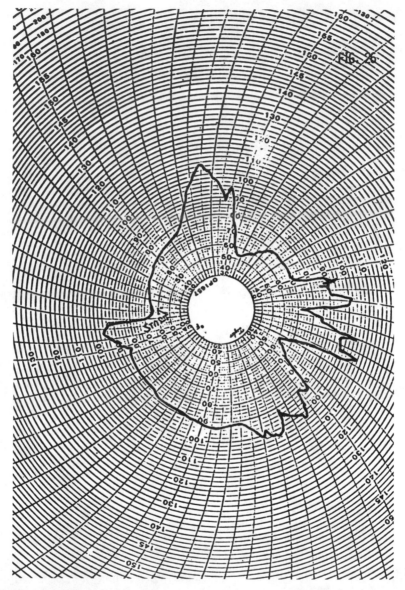

Figure 2.2. Twenty-four-hour recorder chart of effluent temperature on a typical day after conservation practices were put into effect.

Understanding the Problem

The following phrase is as overworked around the globe as it is in the United States: "you just don't understand my problem." In fact, examination frequently reveals that the problem with the person making the statement is the person doesn't understand the problem.

Isn't it strange that for most production people the word *savings* when applied to wastage is almost never equated to the term *lost money*? As I look at reducing cost in a waste management scheme I'm reminded that, in an industry like food processing that generates profits at about 1% of total sales, every thousand dollars savings made in operations equate to about 100,000 dollars of sales.

How much effort do you think our sales people would make to put on new customers willing to purchase 100,000 dollars of product being manufactured? The answer is obvious.

The best way the food industry can treat its waste is to avoid producing it. Most often a factory's sewage load can be substantially reduced by improving in-house waste abatement methods, which are less expensive and far more productive than end-of-pipe treatment.

Waste Source

To begin with, one has to know *how much* waste is being produced, *where* it comes from, and *what* is its composition. Knowing the "what" and "how much" often provides answers and opportunities for coupling unused resources into existing products.

Monitoring waste continuously and keeping records of flows are important because the information provides data for management to determine whether plant effluents meet local, regional, or national water pollution abatement regulations. The practice could prevent public embarrassment or, if too late, counteract it or even prevent nuisance suits.

Accounting for waste flows on a daily basis supplies supervisors with information that detects material wastage. This is of immediate economic importance. The waste, be it cold water, hot water, food residues, or ingredients like sugar, stabilizers, or detergents, shows up as biological oxygen demand (BOD) or chemical oxygen demand (COD) values in sewage. It makes sense to me for the organization to assign reasonable unit cost values, by regular accounting methods, to different waste streams being discharged to sewers. Plant personnel would be motivated to reduce wastage when they know that they can be graded for waste control in plant operations in terms they all understand: money.

It is a pretty-well-known fact that most food processing plants fail to measure processing wastes as a separate identifiable product. Management tends to recognize product recovery in operations only in terms of finished goods either as yield in percentages or pounds recovered per hundredweight of raw material. The real truth of the matter is that if you see product or water on floors of a reasonably modern food factory, it's pretty obvious to "an experienced eye" that the operation is probably suffering from food product anemia. It is discouraging to find that a packaging line is foolishly malfunctioning at startup and losing product in the food handling system after spending valuable time in assembling and sanitizing the equipment.

Of at least equal importance in controlling waste at startup would be the fact that post-production shutdown also needs special instructions because product left in equipment is valuable and has to be drained or rinsed from the machinery before washing operations. The rinsing ought to be diverted to tanks before going to floor drains to be collected as useable product and segregated into different categories such as unflavored or flavored materials. Renovation and reclamation of food waste can be the industry's choice to convert waste treatment costs into economic credits. Besides the monetary drain for waste-handling services, the industry should understand that its basic resources like water and chemicals have definite limits. It is important that companies develop and use more efficient recovery schemes that reduce water usage and sewage charges, which are going to rise substantially in the years ahead.

Cow Water

I demonstrated a way to recover hot water from waste during the early fifties. It involved collecting hot milk vapors as water from condensing evaporators. This accumulated water was later used for boiler feed needs and plant wash water. I coined the term "cow water" for the process. The water temperature of the salvaged fluid generally ranged between 120°F and 170°F. Water quality was both monitored and controlled by passing the liquid through turbidity detectors. Water clarity was chosen as an index of quality. Changes in clarity triggered an instrument that activated three-way air valves, which funneled good water to storage areas and spoiled water to sewer drains. Water spoilage occurred when the evaporator malfunctioned and caused product entrainment losses to become mixed into condensate. Faulty air leaks in the system also created situations or conditions where milk would be carried over in condensing vapors into cooling water.

The method has since been adopted and used by many food plants both in the States and abroad. Swedish workers improved the "cow water" idea of recovering hot water by installing special anionic exchange units to treat recoverable evaporator condensate. The purpose for their innovation was to provide a mechanism that would completely eliminate proteins and volatile acids from gas bubbles in milk or whey vapors and remove burnt particles coming from vapor recompression nozzles. These defects are not readily visible to the naked eye or to a scanning turbidity detector.

Water Quality

Industry people are going to have to think in terms of improving "waste quality," the quality part being defined in terms of pH, COD, BOD, hexane soluble, etc. Some of you already know this because you are probably paying large sums of money for municipal sewage charges and see little to no relief in sight.

The basic rule to reduce both volume and waste strength in effluents discharged from food plants *should be* that *all processing equipment* has to be checked before startup for "fluid tightness". This can be done by pumping tepid water under pressure through the system to check for leaks. (Cold water is not used because the equipment being tested tends to sweat.) By completing this step before starting up with product, the staff can check heat exchangers or packaging machinery, etc. to see if equipment operates mechanically well and to confirm that auxiliary lines and pumps are connected together correctly.

It is no trick at all to reduce total waste flow volumes in most plants by at least 50 percent. Often all one has to do is to go around a factory shutting off running water hoses or repair leaky valves. Note the reduction in water use made in a Brooklyn, New York food plant in a 6-month period as shown in Figure 2.3.

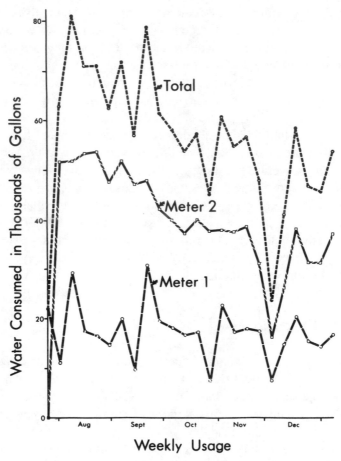

Figure 2.3. Reduction in water use in a 6-month period made in a Brooklyn, New York food plant using water conservation practices.

Accounting Techniques

If we assign unit values to effluents, accounting officers need at least two pieces of information to quantify a plant's liquid waste into pounds of BOD. These are volume and strength. Daily volumes can generally be determined by simply measuring water consumption in an operation by reading water meters at set times, much like accounting for gas or electric use. Other methods might have to take the form of measuring flows through flumes and weirs as required by specific site needs. Weirs and flumes are described in detail in Chapter 3.

The strength information comes from developing a meaningful composite waste sample that, when tested, truly reflects product losses sent to sewer drains. There are many tricks to accomplish this task with sampling needs ranging from fully manual tasks to using mostly automatic systems. Sampling frequencies, sampling size according to flow rate, and sample storage during collection are but some factors in a scheme one has to think about before establishing waste-sampling procedures. Regardless of the method selected by personnel in food factories to build waste composites, the effort will only be meaningful when *the game plan selected for the job is sound.*

Losses can be expressed as:

- Waste Equivalent of Raw Product
- Waste Equivalent of Finished Goods
- Utility Value; i.e., Water
- Utility Value of Power; i.e., Steam/Gas/Electricity
- Solid Waste Disposal Costs

The system selected needs to address the special monitoring details to account for information required.

I'm convinced that plant workers, as well as their managers, have to know that water and waste are real commodities like other raw materials to be measured regularly, day to day, in plant operations. Part of these costs of doing business is fixed expenditures and other amounts are variable. People ought to be able to change portions of variable costs when they know and understand cause-and-effect relationships of their actions.

Fixing poorly-running machinery is certainly a variable.

CHAPTER 3

In-Plant Monitoring

At least two pieces of information are needed to quantify waste loss. These are mass and mass concentration.

For liquid wastes which may include fruits, syrups, product rinsing, etc. we can either contain the liquids being lost or obtain a representative sample to be analyzed for strength and to determine the loss. Most often containment of liquid loss volumes in a single tank would be an unmanageable exercise; therefore, we move to a variety of acceptable measuring schemes.

Estimating Wastage

Flow measurements can be straightforward once site drainage arrangement is put in order. If effluent flow is nearly constant, then multiplying flow times rate of flow will provide mass.

Unfortunately, most food plant waste flows are variable as to type and amount but reflect the variety of unit operation taking place over time.

One technique could be that when waste is sent to a sump the operator records the time a pump of known capacity is online. A number of waste pumps in a large factory can be timed and their output summarized to yield flow.

The more common system though would be to measure flow with weirs. A weir is a damlike structure. Flow is restricted and height of flow above a discharge outlet is a measure of volume passing over or through the dam.

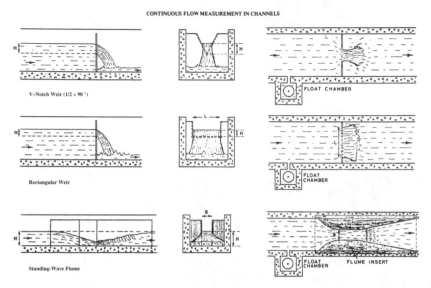

CONTINUOUS FLOW MEASUREMENT IN CHANNELS

Figure 3.1. Continuous flow measuring systems where H refers to height, L is length, and B takes into account throat size. These measurements are used to gauge flow volume.

Weirs can be V notch, rectangular, and Parshall flumes (sometimes called standing wave flumes). Discharge notches of V varieties vary from 22 1/2 degrees, 45 degrees, 60 degrees, and 90 degrees depending on anticipated volumes. Figure 3.1 depicts flow-measuring systems. One tries to size a notch to cover low to highest flow. The best-fit notch delivers the more accurate results. With use of weirs there are installation ground rules to promote accuracy. Hydraulic tables and other data are readily available in booklet form, easily available from equipment suppliers, colleges, etc. The term commonly used to deal with these kinds of data-collecting activities is "open channel flow measurements". There are tables that will show flow in gallons versus crest height for the full range of damlike devices. Figures 3.2 and 3.3 show flumes and Figure 3.4 a discharge curve.

Accurate flow readings are important to quantify waste and loss to a factory. Besides the mass or flow sent down the drain, managers need to be able to sample the flow to determine waste loss concentration.

Sampling Waste Streams

Perhaps the best sampling method would be to contain waste in a single vessel, agitate the mass, then collect a well-mixed representa-

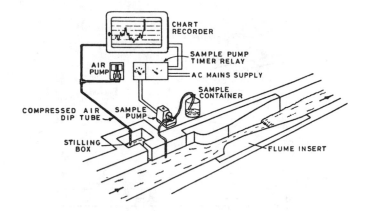

FLOW MEASUREMENT AND SAMPLING USING A RECTANGULAR
STANDING–WAVE FLUME (PERIODIC FLOW DEPENDENT SAMPLING)

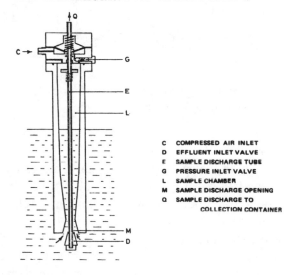

C   COMPRESSED AIR INLET
D   EFFLUENT INLET VALVE
E   SAMPLE DISCHARGE TUBE
G   PRESSURE INLET VALVE
L   SAMPLE CHAMBER
M   SAMPLE DISCHARGE OPENING
Q   SAMPLE DISCHARGE TO
        COLLECTION CONTAINER

SAMPLING DEVICE FOR 'TIME
AND FLOW DEPENDENT' SAMPLING

Figure 3.2. Standing wave flume, sometimes called a Parshall flume
with a sampling device.

tive portion of the waste and use the sample to determine waste concentration.

Of course, when flow volumes exceed ten thousand gallons, this approach is not practical. Sampling choice comes down to spot sampling
or "a grab sample," a periodic sample, and/or a continuous sample.

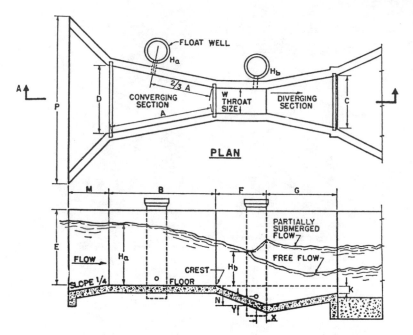

Figure 3.3. Detail diagram of a wave flume. Each of the notations must conform to some rule when constructed to assure measurement accuracy.

Let's deal with the spot sample and get it out of the way. Spot sampling is a random event and will not reflect a true picture of flow and must not be looked upon to provide a cumulative reliable expression of wastage content.

Periodic sampling is time, flow, and time and flow dependent. When samples are collected and commingled to produce a composite sampling the results could be high, low, or about correct. For example, if waste flow is sampled hourly over a 24-hour day and the factory runs product 8 hours, then samples (probably clean water flow) of the same size collected for 16 hours dilute the composite if flow rate is not considered.

Proper sampling equipment is a unit where both sampling time and sample size can be varied. The demand is that each fraction of a sample that makes up a composite sampling must vary sample size according to flow volume. Therefore, continuous samplings are recognized as being time and flow dependent. Figure 3.5 illustrates different types of sampling devices.

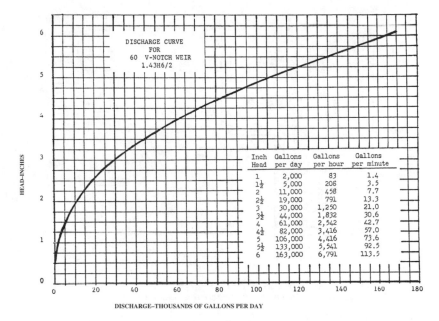

Figure 3.4. Discharge curve for a 60-degree V-notch weir where height measurements in inches convert to gallons per minute.

Do Not Fool Yourselves

A manager who uses too many shortcuts when sampling measuring stations can develop erroneous information. Another caution is that when laboratory staffs test poor samples they too waste time and money by analyzing samples with little to no meaning. No matter how proficient a laboratory worker can be, no matter how many decimal places the analyst puts in the data, the answer has little value.

Solid Wastes Need Attention

Solid wastes can be combined and containerized. The preferred method is to weigh containers and then weigh the empty containers to determine mass. Many times these solids contain food fragments, damaged packaged goods, and a mix of raw materials, paper, plastics, etc.

The only way to approximate composition would be to grind mass and sample homogeneous material. For the most part this activity is too

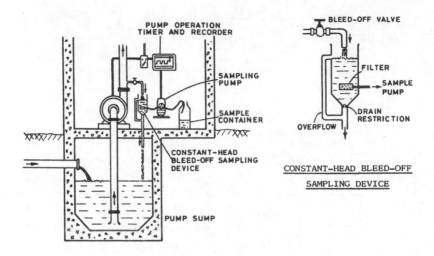

PUMP OPERATION
TIMER AND RECORDER

SAMPLING
PUMP

SAMPLE
CONTAINER

CONSTANT-HEAD
BLEED-OFF SAMPLING
DEVICE

PUMP SUMP

BLEED-OFF VALVE

FILTER

SAMPLE
PUMP

DRAIN
RESTRICTION

OVERFLOW

CONSTANT-HEAD BLEED-OFF

SAMPLING DEVICE

**MEASUREMENT AND SAMPLING OF A PUMPED EFFLUENT
DISCHARGE (CONTINUOUS 'TIME DEPENDENT' SAMPLING)**

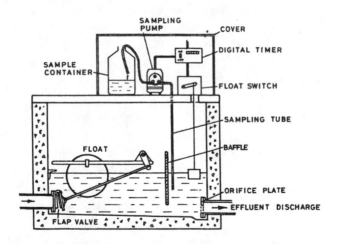

SAMPLING
PUMP

COVER

DIGITAL TIMER

SAMPLE
CONTAINER

FLOAT SWITCH

SAMPLING TUBE

FLOAT

BAFFLE

ORIFICE PLATE

EFFLUENT DISCHARGE

FLAP VALVE

**MEASUREMENT AND SAMPLING OF THE EFFLUENT DISCHARGE FROM A
CONSTANT–FLOW MODULE (CONTINUOUS 'TIME DEPENDENT' SAMPLING)**

Figure 3.5. Sampling arrangements that can be utilized to produce reliable composite samples of waste effluent.

expensive and would not be done on a daily basis. A reasonable solution is to periodically commutate the refuse from an entire shift or even a day's load and obtain representative samples that can be analyzed. This food value, combined with the other mass data, can be used to approximate the dollar value of refuse being discarded.

Of course the waste being generated in a food plant will not originate from a single unit operation. For the most part food waste occurs in fragments or chunks, depending on the food commodity being processed.

During the transformation of raw materials into finished packaged goods, a series of steps take place. Loss of some magnitude occurs at each step. To quantify loss at only the discharge pipe of a factory neglects the ability to pinpoint loss to specific operations.

We ought to think in terms of being able to track product loss at individual work stations. If not at individual work stations, we certainly ought to be able to quantify waste at different key work zones within a factory, such as raw loss/processing loss/packaging loss/cleanup loss, etc. In order to do this we need to diagram drainage areas within the factory. If you accept the idea that it's too late now that a factory is built to relocate drain lines to be able to install monitoring stations, the game is lost before you begin.

If you accept the idea that all drainage systems have to be underground, you have again lost the game before you begin.

Some food plants first drain individual unit operations to a sump within a processing station, then pump waste to a central waste treatment area or factory site.

Aboveground piping is easier to instrument to gather flow volume information. Aboveground installations are generally easier to instrument for sampling waste flow components for analytical purposes.

This does not mean that the same kind of information cannot be generated from in-floor plumbing too.

The fact is that material sent to drains or rubbish bins can be products of some value. The value may be positive as a salable product that you choose not or wish not to sell, or certainly a cost item or negative value that detracts from profits in the operation of a food factory.

Waste audits ought to be and can be a process engineering technique, along with quality audit activities.

The audit not only quantifies loss of food fragments, but also adds to measurements of product-control effectiveness. On a day-to-day basis loss accounting pinpoints or tracks effectiveness of different factory controls, which include raw material quality or lack of it when processing

spins off excessive organics to drains. For most firms, production capacity versus time is surely of such magnitude that when processing
constraints are ignored both product quality and product loss become
excessive.

We may well be overworking terms like "total quality control" or
"HACCP" (hazard analysis and critical control point) in the workplace.
Over time these ideas, as good and as glowing as they are, begin to lose
luster. What does not change over time is that, when people go to banks
to cash checks or make withdrawals, individuals on both sides of the
cashier cage count the money at least once.

CHAPTER 4

How to Carry Out a Management Scheme

A t least four broad options are available to food manufacturing operators to deal with waste disposal.

First, the processor may be able to reduce waste from select process production points so as to generate less waste.

A second choice looks at waste streams as food by-products. These food fractions might find use as raw material for manufacturing new products. In fact, the management looks to the new product as a source of revenue.

Thirdly, waste might be looked upon as a fuel source to burn directly or perhaps be substrate material from which to make alcohol or methane.

The fourth choice is to treat the waste as sewage and go about getting rid of it with the least expense.

Troubleshooting

The basic steps to troubleshoot waste begin with mapping the drainage system for each unit operation. Engineering staff, which can consist of one person, diagrams drains, equipment location, and exit piping.

The next step would be to select waste-monitoring stations where liquid wastes can be quantified. While this task may be troublesome for older plants versus newer operations, the task is not without solution. One firm, the Ocean Spray Company in Massachusetts, uses aboveground drain pipes and pumping systems to move wastewater from its plant to an on-site treatment plant. Small pits were put in strategic locations to collect liquid wastes, which were then pumped. In fact, the aboveground piping makes it easier to install measuring/monitoring devices to track wastage.

Once the monitoring sites are in place, select and install off-the-shelf hardware that quantifies flow as to volume and concentration. You will find a variety of BOD/pH/temperature and conductivity probes are available which can be used to provide useful information.

Once the two initial tasks are completed, data can be collected by using electronic or mechanical collecting systems so that supervisors can keep track of waste and manage it.

Flow and concentration data need to be sent on a daily basis to Accounting, where current dollar equivalent values can be assigned to each loss component. Over time a normal distribution of loss costs will be measured. This does not mean typical day-to-day losses need to be accepted. In fact, what will be set in motion is a series of trend values that depict operating-day losses. Supervisors are expected to be supplied loss-cost data promptly so proactive management can occur. In other words, someone has to look at data being generated and to take action to keep loss figures at a minimum. Wastage monitoring systems make up a series of controls that can be used to alert factory people that changes need to be made in processing technology. Loss can be due to variable raw feed stock, poor quality raw material, faulty machinery, and other problems. Managers will get used to looking at loss figures like caution lights and even stop lights as the occasion demands.

Solid Waste

Solid waste generated in food factories can be quantified using conventional methods. For packaging rooms where losses might occur because of an overweight, jammed container, machine breakdown and other reasons, workforce people need to account for container usage. Packaging losses can be determined by comparing empty containers withdrawn from the warehouse versus full containers sent to storage. Container losses alone do not tell the full story as jammed cups lose different amounts of product. Cup accounting may at best cause supervisors to tighten up on operational practices that vary from improvement in maintenance to finding flaws in packing materials sent to the firm by suppliers. Monitoring systems that can be timely and informative supply proactive information rather than historical data. It's always good to know how things are running, as opposed to how they are supposed to be running. Waste inventories have a way of raising red flags to tell people something is wrong.

My scheme for managing waste is geared to forcing factory managers to accept accounting practices that include water, power, and utilities, as well as product use per shift/per day. Along with the other variety of production inputs managers keep track of—like labor use, overtime hours and ability to meet production quotas—there is value (sense to make cents) in tracking losses sent to drains in key production zones. These losses are part of the combined loss sent as sewage from the entire factory.

Guessing versus Knowing

It's disappointing to find that management people in food plants are in the dark about how much fuel, power or even money it takes to make individual products. I find only a few mid-management individuals have any feel for the amount of electricity, steam, and water consumed, or waste generated in their factory to make different products. Most plant managers know what raw material costs, but few supervisors are familiar enough with operations to know individual costs, like utility expenses, or even labor hours per unit product made.

Obviously, accounting officers need, and sometimes get, good to poor information upon which to build product pricing. Not so obvious to top executives is the fact that their on-floor managers need, but seldom get, processing cost information so they can begin thinking about how to take advantage of emerging food processing technology. What can be done?

If the firm uses cost centers, then you already have the operations divided into different categories. Cost centers usually have a point of origin and end at some point based on how individual management looks at in-house manipulations. If your company doesn't follow this practice, you might consider using a similar system where specific unit operations can be separated from overall factory operations.

Someone has to be given the task of determining just how much electricity is needed, how much steam is expended, how much water is used, and what amount of refrigeration is being utilized. In addition,

you ought to be able to quantify waste losses from each unit operation
to minimize product losses, as well as to save sewage treatment expenses. Some firms do this on a regular basis. Some people know they
have to measure all inputs going to an individual product every time
changing machinery or ingredients modify an operation.

How to Do the Job

Assign the work to a person who will take time collecting accurate
data. In the past, I've hired graduate students to do the job for me during summer vacation periods, and this arrangement worked out nicely.

To measure labor input, the old stopwatch method is straightforward
and easy to eyeball. Operation time is monitored; units produced are
counted and divided by time for appropriate units per hour. Electricity
being used can be measured by taking amp load readings on motors
used in operations. The individual finds out how much energy the machine consumes when in operation and multiplies voltage by amps to
get watts, and watts by time to get energy consumed.

Steam usage can be measured by actually weighing condensate leaving heating systems.

This is often done by simply unscrewing a convenient joint in a line
and then collecting condensate with buckets. This volume, when measured for specific time periods, will indicate pounds of steam used.

Refrigeration probably takes a little more pencil-pushing. The prod-
uct weight multiplied by the temperature difference to be cooled gives
British thermal unit (BTU) figures. The total BTUs divided by 144
BTUs (this is in a pound of ice) gives pounds of refrigeration used.
Fudge the figures with efficiency values, which would probably run
about 60–70 percent, and you would have a good ballpark value.

Sometimes it's convenient to wire electric clocks to magnetic boxes
so they run when refrigeration compressors operate. In 24-hour periods, the plant staff can know the operation period compressors run for
calculating the amount of power used to run refrigeration systems. Obviously, automatic recording devices are available to perform these
tasks.

There is no quarrel with executives who want to keep individual
plant processing costs confidential so they can optimize individual marketing strategy. We can disagree, however, with practices that tend to
dull employee initiative. Better-informed employees tend to take cor-

rective measures to operate more efficiently, especially when they know their activities can be measured. You need to wonder when was the last time you or anyone else in your company did an audit of full inputs and outputs used to produce individual packages.

Dollar Value

Accounting personnel must be given sufficient information from which they can assign dollar values to losses such as solid and liquid waste. Think of waste in terms of product equivalents where organic composition is measurable as BOD or COD and then priced at real wholesale value. Differentiate loss as raw material from finished processed foods, as dollar equivalent values are different. Loss of equal amounts in raw material is worth fewer dollars than as finished product. The work force understands differences, and must not be provided faulty or poor information.

Information obtained from a final monitoring station that reflects combined loss only tells you what total amount of waste needs treatment. While total pounds of BOD are a cost item, it differs from a product equivalent item, as sewage cannot be intercepted for reuse, recycling, etc.

Loss Centers versus Profit Centers

Managers ought to initiate practices that deal with loss centers much like they do with profit centers. In this case the profits are negatives. We might think in terms of loss centers as a conventional accounting system that deals with profit centers.

Sometimes factory waste can be partially treated prior to being sent to a municipal waste treatment system, making treatment a variable-cost item.

Some might think the suggestion put forth is not realistic and even somewhat idealistic. Such an assessment would be inaccurate because the concept described is little more than common sense and well within the industrial means to implement. More sophisticated managers will jump on this kind of a program if motivated to make the effort. What better motive is there to give the work force than that without profits there can be no employee benefits?

Waste Discipline Is Everyone's Business

A novel adjunct to in-house training of supervisory personnel by one English food company is teaching its management trainees how to collect and develop mass balance information of materials entering and leaving its factories.

While the idea of teaching production people the importance of quantifying both amount and composition of incoming receipts is usually stressed by most companies, the wastage end of materials from manufacturing processes is almost always ignored or forgotten.

The trainee in one industry is given the assignment to follow food and service material through a factory and account for their deployment in its different unit operations. Material for audit is taken to mean readily measurable items such as flour, milk ingredients, sugar, salt, cleaning liquids, water and sewage. Mostly work-study people collect the stream and electricity usage data from the firm's engineering department.

As you might suspect, the individual making a mass balance chart must learn how to collect or develop representative samples that portray different levels of production. Following this step, he or she must then go about quantifying the flow volumes by composition in order to generate meaningful and useable proximate analysis data. The learning technique appears to be a surefire way to teach management people to remember their training lessons. One plant manager who had already gone through the program remarked, "Waste truly becomes a real identity to you when you personally measured the stuff by the bucket."

The former trainee who by now had risen to a manager position himself and could judge the merits of his previous training was now adding his own innovations to training personnel. He developed a system that he called "waste discipline" and used it for controlling product losses in his own factory by involving all plant people directly with the operation of the factory's sewage plant. From time to time, factory workers are required to tour the company's waste treatment facilities. While there, they are brought up to date with the nut and bolt problems of treating waste. Production people soon learn to recognize the fact that the sewage part of their plant is an important department that needs attention like other equipment in the factory. This service department is needed by the company to keep the factory running.

Supervisory personnel are urged to visit the treatment plant regularly. They learn by experience that processing errors such as lost products from the different operations within the factory can easily be pinpointed in the waste plant. Telltale signs such as food fines appear on incoming mesh screens; fat balls arise from lipid operations, oily cream from separation. Different people in the plant are now using the term "waste discipline" because they are convinced that they can manage waste.

From their point of view, accidental production problems, if and when they occur, should not be allowed to complicate matters or to cause further damage to the firm by upsetting or shocking the efficiency of its waste treatment works. Plant people feel they are able to control conditions in the sewage plant when accidental spills are known by making appropriate manipulations to hydraulic flow rates to help the system treat sewage properly, even under adverse situations.

CHAPTER 5

Product Loss and Dollar Equivalents

Food processing wastes range broadly depending on commodity. Wastes from fruit and vegetable processing plants differ from wastage from meat slaughtering, as do the kind of discharges from poultry processing plant wastes. The food operations can include milk processing operations and seafood wastes as well as a myriad of products from food specialty factories. What is common to the different wastes from food operations is that we know the liquid wastes would probably be made up of at least moisture (water), carbohydrates, proteins, and fat. There's a plethora of information available in tables of different textbooks that provide characteristics of food processing wastes. These data are listed as COD, BOD/COD ratios, pH, etc. and provide mean or average loss values per gal/case of can products or other convenient measure common to the industry. Our Environmental Protection Agency (EPA) compiled the more recent information when they funded projects that looked at waste from different commodities. My research group provided the EPA project reports with factory processing information about waste profiles concerning dairy, egg-breaking, and seafood operations.

These tables, etc. have been helpful to civil engineers given the task to build waste treatment systems for different food processing factories. However, the best treatment systems still originate from data generated from individual plants by assessing the specific operations wastage. BOD/COD values in waste broadly measure the combined gross composition, not individual components, and will give the best information to a processor to pinpoint exact losses from different unit operations.

Recommendations

What is being recommended is that multi-station sampling points from different unit operations gather composite samples of representative flows discharged per shift in food processing areas for analytical purposes. These samples need to be tested for COD or BOD, plus those components specific to the foodstuff being processed. If we assume the raw material produced into a food contains carbohydrate, fat, proteins, and minerals besides water, then the different components have variable values depending on the ingredient source at the sampling station. Obviously, raw material components would be less valuable than components lost in products at other stages prior to the finished good value.

For example, product X might be worth 20 cents/kilogram as raw product when the worth of components with carbohydrates is 5 cents/kilogram, 15 cents/kilogram for the protein fraction and 10 cents/kilogram for fat content. Further up-line in processing, when product is cooled and concentrated, the protein and fat content is increased. In fact, the kilogram value might double to 30 cents for protein and 20 cents for fat losses with little change in carbohydrate value. When product loss occurs at this step more loss value needs to be applied to components sent to floor drains. This kind of an evaluation process quickly shows that a firm needs to begin paying strict attention at key steps to maximize product yield. All losses are vital, but we are perfectly safe in thinking more money will be lost in finished goods or partially finished products than for raw material in a food processing sequential step operation.

Here are the things that need to be done:

1. Establish a loss sampling station.
2. Gather and test representative samples of loss materials sent to drains.
3. Assign dollar values to different components.
4. Provide unit operation supervisors loss information promptly so corrective actions might be taken to improve operations. On the other hand, the data may well confirm to people that the operation is being carried on with success. This too is vital for people to know.
5. Inventory wastage with the same vigor you would do for other items in your shift or day accounting system.

Waste Monitoring and Cost Control

One reason for monitoring waste and keeping records of flows is that the information provides data for management to determine whether plant effluents meet local, state, or federal regulations. The practice could prevent public embarrassment or, if too late, counteract it or even prevent nuisance suits.

Accounting for waste flows on a daily basis supplies supervisors with information that detects material wastage. This is of immediate economic importance. The waste, be it cold water, hot water, food residues or ingredients like sugar, stabilizers, or detergents, show up as BOD or COD values in sewage. It makes sense to me that plant personnel assign reasonable cost unit values by regular accounting methods to different waste components being discharged to sewers. The work force would be motivated to reduce wastage because they would know that waste in plant operations could be expressed in terms they all understand: money.

For example, suppose a plant decides to measure the following parameters on a daily basis: total flow, temperature, and BOD. The accounting manager establishes these cost values: water is worth 80 cents/1000 gallons; fuel is valued at $8.00/million BTU; solids are worth 50 cents/pound; flow amounts to 100,000 gallons; average temperature is 75°F where raw water is normally 55°F; and BOD concentration in waste composites is found to be 400 ppm.

Calculations:

Water loss value

$$100,000 \text{ gallons} \times 80 \text{ cents}/1000 \text{ gallons} = \$80.00/\text{BTU loss (approximate)}$$

$$100,000 \text{ gallons} \times 8.34 \text{ pounds/gallon} = 834,000 \text{ pounds}$$

$$\text{Temperature rise/ gallon of water} = 75°F - 55°F = 20°F$$

$$\text{A BTU} = \text{a pound mass raised } 1°F$$

$$834,000 \text{ pounds} \times 20 = 17 \text{ million BTU}$$

Or

$$17 \text{ million BTU} \times \$8.00 = \$136.00$$

Note: Refrigeration losses such as cooling water could have cooled waste flows and masked some heat loss.

Pounds of BOD

$$\text{Pounds BOD} = \frac{\text{Gallons flow} \times \text{wt/gal} \times \text{BOD ppm concentration}}{1,000,000}$$

$$\text{Pounds BOD} = 340 \text{ pounds of BOD}$$

In general a pound of food waste solids is equivalent to about a half pound of BOD.
  Or

340 pounds BOD $\times$ 2 = equivalent to lost food solids = 640 pounds

Or

680 pounds $\times$ 50 cents/pound = $340.00

The value of the sewage effluent discharged by the plant in this example could be $80.00 + $136.00 + $340.00 = $556.00. If this amount of waste continued as is for a 250-day/year operation, this discharge could amount to $139,000 cost against the plant operations.

Certainly most plant manager personnel would know from these data that the dollar loss or waste cost could be shifted most easily by attacking the pounds of BOD figure. BOD could soon be equated with organic losses, and then these units would logically be equated with ingredients or like commodities. The name of the game is to reduce food organics wasted to drains and decrease total flow.

The responsibility for insisting on daily accounting as a regular survey begins with a firm's senior executive. Plant management with the cooperation of in-plant workers would carry out the task.

Approximating Cost and Lost Pounds of BOD

Two pieces of information are needed to quantify liquid waste into pounds of BOD. These are volume and strength. Daily volumes can generally be determined by simply measuring water consumption in an operation by reading water meters at set times, much like gas- or electric-use accounting. Other methods might have to take the form of measuring flows through flumes and weirs as required by specific site needs.

The next step is developing a meaningful composite waste sample that, when tested, truly reflects product losses, plus water mixtures of products sent to sewer drains. There are a number of tricks to accomplish this task with sampling needs ranging from completing fully manual labor tasks to using mostly automatic systems. Sampling frequencies, sampling size according to flow rate, and sample storage during collection are but some factors in a scheme one has to think about before establishing waste-sampling procedures.

One food plant built its waste-monitoring center so that sewage is segregated into clean and dirty water canals flowing through the two-hatch-covered pits. Metal valves can be manipulated to allow either or both types of waste to enter a flume-measuring device. Figures 5.1, 5.2, and 5.3 show the installation.

The process has the flume equipped with a constantly-running, wastewater pump. The pump feeds an automatic sampling device, which builds a composite sample of the day's flow. This composite sample accumulates in the plant's boiler room where the firemen on duty monitor waste flow, along with steam flow.

Figure 5.1. Waste-monitoring site with flume located inside small building. Two outside covers house control valves where clean and dirty water can be segregated and measured.

Figure 5.2. Parshall flume with thermometer probe inserted into flow. Also shown is pump station.

An automatic device draws samples, the sizes of which vary in proportion to the flow of wastewater. The instrument measures flow and provides a chart record of flow in gallons, temperature, and wastewater BTU content above that of the raw inlet water. Such a monitoring system is an excellent tool for better plant operation.

Once a plant has collected a representative waste sample a rapid, inexpensive, and simple test can be used to determine its strength. A rapid COD test was developed as early as 1967 that takes less than 30 minutes to run and at about one quarter the cost of the Standard Methods Test. The procedure can be used advantageously in most food plant operations, even to help small laboratories to curb plant losses. The method is not intended as a substitute for Standard Methods, but it is sufficiently reproducible as a management tool for day-to-day needs. The procedure was described by Dr. John S. Jeris of Manhattan College, New York, NY, and can be found in *Water and Waste Engineering* Vol.4, No.5, May 1967. Since then improvements in carrying out similar test methods have been developed.

A more recent technique to test BOD is reported by Meggle Milchindustrie, a dairy plant processing 500,000 liters (132,000 gal.) of milk per day at Wasserburg, Germany. The plant dramatically reduced wastewater BOD, reduced product loss, and slashed the electricity costs of waste treatment by 30 percent after installing a BIOX-1000 continuous

Figure 5.3. Instrument/sampling station that charts flow and converts temperature readings to BTU loss above reference point and composite sampling station.

short-time BOD monitor. Payback was achieved within a few months. The monitor detects spills within 3 minutes, and triggers an alarm.

The pH and conductivity measurements integrated with the BOD monitor pinpoint origin and cause of the spill, enabling the plant supervisor to take immediate action. The BIOX-1000 is based on the BOD-M3 method, whereby the computer to control the nutrient concentration in the waste stream uses the respiration rate of the microorganism. The BIOX-1000 continuously measures BOD within a range of 2 to 10,000 mg/l and presents results on a digital display. A printout is generated as electrical signals to communicate with process control systems. The heart of the system, as shown in Figure 5.4, is a bio-reactor containing

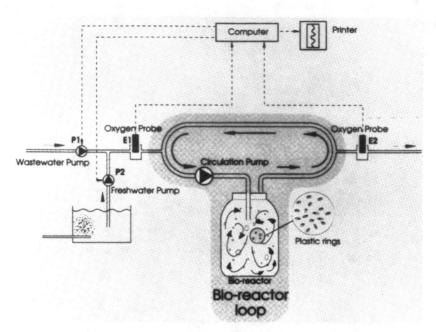

Figure 5.4. BOD monitoring device, which measures biological respiration with speed.

many small plastic rings, which provide growth surfaces for microorganisms and are maintained in turbulent motion by a circulation pump. Wastewater is diluted with potable water to a low but constant food concentration by metering pumps in a control circuit, where oxygen consumption of the microorganisms (as measured by oxygen probes) is the control parameter. The instrument's computer, based on the dilution ratio between wastewater and potable water, calculates BOD.

CHAPTER 6

Improving the System

Improving the system might be thought of as "plugging the drain." While we may not be able to run food plants without waste, there are methods to put control measures into use to reduce wastage.

Be Informed

A running inventory of checks avoids operating the factory without lights or without the use of a speedometer. Benchmarks are needed to grade or assess work conditions or results. We need to test samples (even if only small amounts) for all lots or samples of incoming shipments.

The terms "quality control" and "quality assurance" are not synonymous. In quality assurance programs we ask ourselves the question, "are we doing the right things?" As for the quality control part, we want to find out, "are we doing things right?"

Waste management as a discipline is, in fact, related to quality assurance programs. As such, it assumes still another role in our fight to lessen product losses within the food factory.

Managers everywhere need to reexamine the instruction given their laboratory directors relating to monitoring product quality. It's worth repeating that preventing waste is a better system by far than treating the waste as sewage.

In some industry plants the quickest and probably the least expensive tests that I know about are smell- and taste-related. How foolish it is to see good product being made bad by mixing sweet ingredients with sour. I'm certainly turned off by the all-too-common comment by line workers, "You want me to taste the stuff?" Believe me, a good nose goes a long way in keeping waste down.

A good assurance program also attempts to cull substandard raw material from a factory before accepting it. The program should also prevent people within the factory from using bad goods, no matter where they come from, on those rare occasions when a vendor delivers substandard material.

It is important to monitor flow/concentrate over time so as to describe factory performance. Day-to-day operations performance can be statistically useful with data collection recording reference.

Treating Wastewater

Like it or not, food plants generate waste, and some staff people have to deal with proprietary waste treatment systems or at least with minimizing the costs that firms pay others for cleaning them up. To do a better job in either—or perhaps both—cases, we ought to expect people to understand what takes place.

In general, wastewaters contain pollutants in the form of suspended and dissolved solids. These solids can be inert or biodegradable, and can be removed from water by two basic methods.

Method 1 uses physical techniques with or without added chemicals. In the simplest of situations, wastewater is screened. The process excludes pollutants from liquids by passing wastewater through hardware such as bar racks, mesh screens or even sand filters. If the wastewater contains oil or grease or colloidal material, then we might use a flotation technique where air is injected into a waste stream and the waste particles are trapped in air bubbles, which float to the surface and then are removed mechanically.

Directly opposite the flotation concept is another scheme, which is called sedimentation. The rate at which wastewater flows into a system is slowed enough to allow settling, collection, and removal of solids. You may be familiar with waste hardware equipment called clarifiers. They're not cheap and are used to settle material and to polish water to which flocculating or chemical aids have been added.

Similar physical separation methods are also used in conjunction with the second method of treating wastewater, which is usually biological treatment. This approach, Method 2, is based on the principle that microorganisms use organic (and some inorganic) minerals in waste materials such as food; by consuming these substances, they grow and multiply. These organisms consume the dissolved material (we have already removed most of the suspended material) and replace these solids with cell mass. At the same time, the liquid environment changes as the organisms also produce a whole spectrum of by-products, as one would expect them to do in a fermentation-type reaction. Microbial growth activity is affected by the kinds of waste, the oxygen content, pH, presence of toxic substance, and other factors.

The next task after creating a working biological system is to remove those solids that have been changed from dissolved material into visible and agglomerated biomass. This is done by using many of the physical separation methods cited in Method 1.

Wastewater treatment is frequently broken down into three broad categories:

Primary treatment removes pollutants by settling basins or collection screens. This would be the first step in a waste treatment system. Primary treatment should not be confused with pretreatments where you are asked to adjust pH or cool waste in order to make it amenable to biological treatment. It can remove up to about 75 percent of the waste load in a system.

Secondary treatment almost always uses a biological approach to treat wastes. It can incorporate physical and chemical methods within its system to remove pollutants. This method, utilized along with primary treatment, will clean up pollutants into the 85 percent range.

Tertiary treatment is perhaps our most costly of treatments. Most sewage treatment plants do not provide such treatment, but may be forced to do so as we move into this century. Tertiary treatments remove such pollutants as ammonia, heavy metals and phosphates. To do this job, the wastewater effluent from secondary treatment would be fed into equipment like ion-exchange, activated-carbon columns, and others to remove pollutants that adversely affect the environment.

Wastewater treatment may take different form and style because the design engineer balances such factors as waste type, volume, plant location, and energy use. It's up to managers to help "marry the many factors" so the system works well, lasts a long time, and proves to be economical.

Overkill in Waste Systems

Engineers can overdesign a waste treatment plant because hardened cement and fixed liquid flows set by constant volume pumps and large pipes suggest that waste amounts to be treated in a given system will probably not be less. Such a premise is obviously incorrect, and this fact is evident and more visible in industries with seasonal variations in workloads, such as dairies. Waste loads in food plants often can vary as much as five- to tenfold above average load levels. The changes in factory operating conditions surely try the ingenuity of our sanitary engineers who are trying to design appropriate waste treatment plants to cope with wide fluctuations of wastage that create the operating demands of a system.

It seems to me that food plants can pay an ongoing costly price for operating a waste treatment plant if engineers do not build into their waste system those operational economies that will optimize running costs to bracket both low and peak load situations.

The failure of a firm to plan for realistic variations in its sewage load showed up very clearly in one factory I visited in the U.K. with a new waste treatment plant. It is a classical example of building a system with an "overkill." Basically the working end of the treatment works is its multimillion gallon lagoon, which is both aerated and mixed by 16 thirty-horsepower floating aerators.

The plant was deliberately overdesigned in size by a factor of two so as to duplicate a sister plant's system because the first plant ran very well. In addition, the installation was supposed to handle waste from another operation, which never materialized.

The logical question might then be asked, "Okay, this can happen, so why not shut off some of the agitators to at least save electricity?" (480 HP operating around the clock is the sizable expense item in the running of the plant.)

Not so simple at this point, because a big lagoon—and in this case a 10-million-gallon amount—needs a lot of horsepower just to keep the system fully mixed. Incomplete mixing in a basin usually creates settling conditions, dead pockets, and odors, as well as exhibiting other undesirable spin-offs.

The waste plant can probably be modified by techniques such as quartering the lagoon with weighted curtains. Perhaps fewer aerators might be located in isolated areas to mix segregated sections in the pond to save energy. However, most changes to be made at this stage will most likely be awkward, and at best will probably be less efficient.

The communication problem between parties involved with the work must have been substantial, at least in planning, if not in the building of the waste treatment plant. Engineers and company directors were certainly not in any reasonable agreement as to the amount of product planned to be manufactured. They failed, it would seem, to consider seasonal operating costs, if not on day-to-day situations, especially as load levels diminished.

One would think if factory managers could not project a reasonable spectrum of load levels for long-range operations in a plant, then the services to operate the installation would have to be flexible.

Power Costs and Oxygen Uptake

Considerable amounts of electricity go into running rotors, air pumps, diffusers, and other equipment for putting air into wastewater at treatment plants.

Sometimes the plants don't appear to get the most for their money because of non-optimum operating practices. Oxygen uptake in wastewater is affected by turbulence, solids, and temperature.

Oxygen is only slightly soluble in water temperatures (about 7–12 mg/l; parts/million); thus, more has to be added when carrying sewage. The purpose of supplying oxygen to water in an aerobic process is to provide the bacteria with energy they need for synthesis and respiration. Total oxygen that has to be supplied would be at least the sum of the amount of oxygen needed for synthesis and respiration. Some aerating devices are also designed to provide the mixing action needed in the system for blending microbial mass with waste and oxygen, even though some think mixing and aeration ought to be separated.

In aerobic wastewater treatment systems, dissolved oxygen is continuously being used for microbial metabolism; therefore, it needs constant replenishing. Various constituents in the wastewater oppose the transfer, which can be measured in terms of oxygen transfer coefficients; that is, one experimentally compares the transfer values for wastewater and tap water under similar conditions. A term "alpha" is used as an index of efficiency.

Surface-active agents like those present in detergents affect oxygen transfer. They concentrate on the air-liquid interface and act as a barrier to oxygen diffusion. Reduction in transfer rates up to 70 percent has been shown to occur in some plant situations. Surface-active agents do not only include detergents that might be added to the waste, but also metabolic products or organic matter being degraded in the treatment system.

Mixing is important and cannot be overlooked. It brings microorganisms into direct contact with dissolved material; thus, turbulence is required for biological reactions. With increased mixing, water layers in waste change at faster rates, thereby resulting in higher oxygen transfer action.

In the initial phase of transferring oxygen to a system, at least, it appears that temperature is only important in determining oxygen saturation. Some researchers have shown that the effect of temperature could not be observed in highly turbulent systems. Also, the influence of temperature on the transfer of oxygen does not really apply.

Oxygen uptake rates vary over a wide range depending upon the type of waste being treated. Because flow rates into waste systems are difficult to control, it is unlikely that waste plant operations can keep dissolved oxygen levels constant. Either more oxygen is being supplied than can be used or, at times, not enough is supplied, which soon causes obvious odor problems. On the other hand, excess oxygen input to a system is not usually noticed, and it is difficult to see that some of the energy is simply wasted.

While aerators can be installed in systems with adjustable weirs so oxygenation input is varied automatically, in most they're not. Because it is relatively inexpensive to install reliable dissolved oxygen monitoring and control equipment to keep oxygen levels at optimum levels, it is surprising to find so many plants ignoring this technology.

Value of Interpreting Analytical Data

There's not much sense in having people run samples of waste through the laboratory if no action is taken to use the information or if it's impossible to interpret the results.

Waste treatment plant performance is best assessed by analyzing 24-hour composite samples. Pollutant removal efficiency can be determined by comparing influent data with effluent results. You want to know if you're getting value for money spent and if the plant is doing a good job.

If BOD and COD are both determined on samples of influent wastewater, you can use BOD/COD ratio values to monitor the degree of waste biodegradability. Values of about 0.5 indicate that materials in the waste are biodegradable. As value drops below 0.4, you ought to be alerted to the fact that the characteristics of the waste may be changing to less biodegradable pollutants. In dairy plants, BOD/COD values should be about 0.5 or slightly above.

In approaching solids found in wastewater, the common analysis methods deal with: 1) total solids, 2) total suspended solids, 3) volatile suspended solids, 4) fixed suspended solids/ash, and 5) dissolved solids. Figure 6.1 shows the relation of these different measurements.

Volatile suspended solids and suspended solids values are interpreted to be biomass or the viable portion of sludge. In raw wastewater, suspended solids are particles, which may be complexes of organic and inorganic material. Suspended solids in final water discharge contain organic matter that contributes to oxygen demand and are, therefore, limited in most discharge permits. High suspended solids in final effluents also tell waste treatment personnel that mixed liquor suspended solids are not settling properly; thus, the treatment plant needs attention.

Figure 6.1. Sequence of steps used to determine select solid parameters in wastewater.

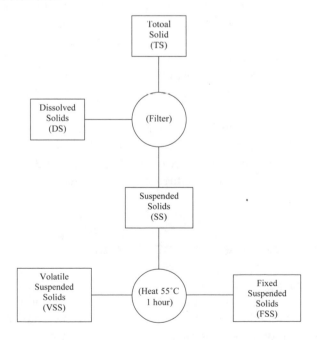

Dissolved solids analysis, on the other hand, is of value to you when the influent contains mostly soluble BOD/pollutants such as those caused by spills from sugar syrups or brine. The chances are pretty good that some kind of dissolved solids additives need to be used to help extract such material from the wastewater. Dissolved solids are not frequently used in the management of a wastewater treatment system, but we may be overlooking its value as a management tool.

Familiarity with commonly used terms to describe waste and wastewater treatment schemes helps in the understanding of this technology. A glossary of terms is provided as an appendix. Some of the not-so-common terms, which may be strange to plant managers, are as follows:

Sludge floc solids are biomass; they are analyzed as volatile suspended solids or suspended solids.

Sludge retention time and *sludge age* are terms used interchangeably. Sludge, by design, is held within waste treatment systems for specific periods according to how a waste system was engineered. The retention time/sludge age is controlled by wasting sludge from the system extracted from some kind of a return line. If, for example, 5 percent of the daily sludge is wasted, then the treatment plant needs 20 days to do a complete turnover in sludge solids.

Sludge settling compaction property is the ability of sludge solids to settle and develop good density. Settling velocity is influenced by the density of suspended particles, and is termed *Sludge Volume Index (SVI)*. SVI is expressed as the volume in milliliters occupied by 1 gram of suspended solids per liter sample size. A lower sludge volume reflects better sludge compaction.

Familiarity with these few terms and others readily available in most textbooks will greatly assist you in controlling waste treatment processes for your food plants. An understanding of the usual parameters can help you maintain satisfactory performance of waste treatment systems.

Automating Treatment Plants

Biological waste treatment systems work best when microorganisms have their nutritional requirements satisfied. To understand how best to manage microbial flora so we can optimize their work as we use them in breaking down organic substrates and/or synthesizing cells, we need to become familiar with the way cells function. In order to keep waste plants running at peak performance, it is necessary to focus attention to the biology of microbial cells. We have to manipulate a variety of separate controls, which together tend to optimize environmental conditions favorable to microbial growth.

Composition

The chemical composition of cells includes carbon; nitrogen; sulfur; phosphorus; oxygen; macronutrient minerals such as potassium and magnesium; and micronutrient minerals such as zinc, cobalt, and molybdenum.

Microbial cells generally contain about 50 percent carbon, 20 percent oxygen, 10–15 percent nitrogen, 8–10 percent hydrogen, 1–3 percent phosphorus, and 0.5–1.5 percent sulfur on a dry basis. The macro- and micronutrient constituents can be measured in the ash portion of cell samples; but, for the most part, chemical content varies depending on growth media.

In general, the need for supplying additional micronutrients to waste material is so low in a treatment plant that it is almost never necessary to add additional amounts of these substances to the system. However, from time to time, macronutrients have to be added to the effluent from factories like those using a demineralizing process, especially when cations are removed from wastewater upstream in a factory before spent fluids are sent to the sewer.

Microorganisms will use different pathways to nourish their growth when synthesizing cell mass. For example, we know that autotrophic groups survive without added sustenance and get energy from sunlight or through the oxidation of inorganic components. Chemotrophs, on the other hand, oxidize inorganic compounds for energy sources and are especially important to treating wastewater such as in the denitrification process. Regardless of whether or not a waste treatment plant is engineered to be aerobic or anaerobic, the chemical reactions that take place in a system obey the normal laws of chemistry and thermodynamics. Microorganisms will use their specialized systems for obeying these laws.

What this information translates into is that in order to keep waste plants running at peak performance, it is necessary to focus attention to the biology of microbial cells. We have to manipulate a variety of separate controls, which together tend to optimize environmental conditions favorable to microbial growth.

CHAPTER 7

Management Tools

Control Charts in Managing Waste

Control charts can be, and are, used to document historical activity in the factory. The canning industry uses retort charts to confirm that batch thermal processes meet appropriate time-temperature protocol to give the correct "D" rates to safeguard canned foods. Of course we cannot do away with installing and operating the large number, often in the hundreds, of charts used in a factory.

The shortcomings of historical-type control charts negate the opportunity by managers to take action when a deviation of norm takes place where the event is only recorded. Some sort of an alarm system needs to be put in place using control charts and allowing proactive reactions to keep a process on target. Just as these needs are vital to our line workers in the factory, controls can be used as process-control techniques in wastewater treatment.

Most managers know about Dr. Edward Deming's work with the statistical process control concept. He can be credited with the idea that the trick in management is to describe complex information into simple clean terms. Just as most control charts record history and not action, he noted that inspection alone did not improve quality, as inspection is just too late. It may be useful to think again about a quote attributed to Deming, which is "A man that does not know his costs, nor whether he can repeat today's distribution of quality, is not a good business partner." As most of us know, we cannot inspect quality into a product, nor can we control waste treatment costs by merely charting flow rates without proactive actions.

Interpretation of Trends

One of the more obvious locations the Food Process Manager looks to assure least loss is in packaging operations. Obviously, no manager worthy of a manager's title can take pride in giving away products via overweight. Early on in the management control charting exercise, the task of plotting package food weights over time was given to filling-room supervisors in order to see whether or not filling machines needed adjustment to keep weights in a targeted range. We all know about operations where charts are now manually or automatically being constructed showing over/under/median/target, etc. The trick of the game is to be able to determine whether or not trends are developing that would cause over- or underfills outside standard levels of acceptance. This being standard operating procedure makes the case that such diligence needs to be given to wastage sent to floors/dumpsters/drains, etc., and losses need to be monitored.

About 20 years ago I published a note about using computers as waste control tools. The article appeared in *Dairy Field,* March 1983 166(3): 68, 69. Since then I'm pleased to see where some organizations are beginning to incorporate such systems into their own plan of work.

Computers as Waste-Control Tools

One of my friends who keeps tabs on plant losses for a major dairy/food company used an Apple II computer to reduce waste and save money. Losses are identified, quantified, given dollar values, and then stored in memory bank disks like those being used with different brands of inexpensive computers (inexpensive meaning a few hundred dollars).

Programs put onto floppy disks are used to allocate losses directly to unit operations and/or distribution points responsible for causing losses. Computer printout information is sent to plant supervisors or even individual machine operators containing data that show how much product or money was wasted due to some malfunction or breakdown in quality control.

Within minutes, pertinent information can be accumulated, put in tabular form, analyzed, and placed into the hands of people who can take corrective action to eliminate problems. The important advantage to using a tool like this is that people who actually handle or make the food products will be shown that their work is being graded and defects can be identified. Such visibility normally improves both work habits and productivity.

A new use for putting tabletop computers into the hands of plant people appears to be for monitoring returns. In this case, the route supervisor categorizes returns as to type/unit/number of units/cost/point of origin defect, if any, by code. Defects such as pinholes in blow-molded jugs, out of code, loose lids on foods like cheese, etc, are coded with some number, and then the code is used for keeping running inventories.

Depending on how a program is put together, it's both quick and simple to produce sheets of information containing data by category. It's also possible to get a handle on data as accumulated to see whether or not some group of defects may be becoming statistically significant. This analytical technique can red-flag more serious problems. Information so collected can be relayed electronically to different plants where corrective actions can be taken to stop production malfunctions.

The rapidity with which information can be assembled using computers overcomes and addresses complaints that supervisors raise about having to wait too long for the cost accounting department's monthly statements to tell them they "had" a problem. The use of computers to monitor returns is an ideal tool for making on-line changes to eliminate or at least minimize troublesome situations.

Almost anybody can be shown how to operate a computer. Within a couple of days, he or she will be able to punch away at a keyboard with or without a display screen like a pro and start using these gadgets for the aid they're meant to provide. I'm told the dollar return on using computers in the firm cited for activities like the one described was fantastic.

Elsewhere in food plants, we are seeing computers being combined with microprocessors to fine-tune thermal or cooling processes. In addition, these low-cost units are now helping managers improve processing efficiency. "Software" programs developed by computer corporations can be used to standardize ingredient composition in continuous flows or to help select formulations for maximizing quality or minimizing costs.

While it's pretty much "old hat" to use microprocessors with computers to adjust electricity/steam/water, etc. based on need, using head pressure temperature and other sensor devices within a factory, there are still lots of room to use these parameters and others to reduce waste. Photoelectric or conductivity cells can be used to monitor liquids in condensate or waste lines; the data, when integrated into computers, can be programmed to take any number of action steps to help manage factories. The truth of the matter is that it's not very fashionable to be intimidated by computers. These "toys" can bring plant people real pleasure and success in waste-reducing areas.

Moving quickly ahead from floppy disks of the 1980s, we ought not forget about our 1990 technologies with video and fax systems. As recently as the spring of 1992 I spent time with an organization in Australia that thinks it's useful to put video camera systems (which are really inexpensive) into sewage distribution manholes of a butter oil factory to monitor wastage. Figures 7.1 and 7.2 show the sump and video camera. Film is recorded on a VCR below the TV screen in Figure 7.2. The color

Figure 7.1. Manhole showing discharge lines from a butter-oil-processing area. A color video camera is mounted in the sump area to record flow appearance.

Figure 7.2. Video camera, TV monitoring screen, and VCR to keep record of film. The video camera is normally installed in the sump shown in Figure 7.1.

camera photographs flow and transmits pictures back to a central waste supervisor's control point. Should visible abnormality in flow liquid occur, the idea is to alert production. At the same time a tape is kept to document the event.

In addition to keeping a watchful eye on wastage should flow become abnormal, data are collected to monitor specific product-loss parameters. In this operation, management tests samples from waste composites for fat and protein content besides COD values. The information is plotted at appropriate locations on flow charts collected over time at different locations. These data are further analyzed for lost dollars; the information is passed on to shift supervisors, informing them when the losses occurred. See Figure 7.3.

As you might suspect, some supervisors are not pleased with the systems, but for the wrong reasons. In this factory top management failed to explain the concept adequately, so the system looks to shift supervisors

Figure 7.3. Waste manager holding a video housing that is put into manhole sump area.

like "big brother" is looking over their shoulders. The waste manager who was not given supervisory status was not getting the kind of coop- eration he thinks was needed. Failure was not from line-working em- ployees, but by mid-management staff to minimize waste. A factory that generates unused by-products really needs some office or person to keep waste under control. From my point of view, food factories need waste watchers. Figures 7.3, 7.4, and 7.5 show a waste manager, an office, and select charts used to monitor waste.

Waste Warden

It's no longer possible for a single competent individual to operate a factory as manager without sharing the mantle of responsibilities. While we recognize this fact to be self-evident, we fail miserably to

Figure 7.4. Waste management office with TV monitor and select graphs used to control effluent quantity and quality.

understand that food processing operations need a waste warden just as they do any other line supervisor.

Factory executives should establish a *full-time waste warden* position in their table of organization. Give this individual a badge, a special hat, even a club with some real authority, and then pick the person who will have enough teeth to bite as the need arises. Such an individual earns his keep in a company by converting waste to profits and provides the firm with peace of mind by keeping the plant out of waste-disposal trouble.

The trick to the whole waste-control program is, of course, control. This means full control every day and not just on selected days. A waste-control person should be responsible to the manager and should not be given other responsibilities, such as those of relief foreman or vacation replacement supervisor. A company that appoints someone to be in charge of its waste-control program has to make the commitment that this assignment is a full-time position. We have already shown that

Figure 7.5. Product constituents plotted with waste effluent volume over time.

surveillance and prompt action are the essential keys needed in operations to avoid so-called phantom losses.

A free-wheeling, high-flying manager, no matter how good the individual might be, is not around often enough on the work floor where supervisors need to show leadership like directing on-the-spot repair of leaky pumps, valves or homogenizers. We are all familiar with wishy-washy kinds of supervisors who would rather rinse product losses down drains than fix them right then and there on the spot. These people rationalize their actions by believing that the plant would lose more products by fixing a drip. In the short run it may be true, but not so over a period of time. When we see food constituents or water on plant floors, it's pretty obvious to an "experienced eye" that the operation is not being well-managed. There is no reason why anyone in a factory should be up to his or her armpits in water or spilled products. The state of the art in processing equipment is sufficiently developed to keep products out of sight during manufacturing—so much so, in fact, that visitors may well wonder what commodity is processed. This situation occurs in properly designed and operated factories.

In recent years, a number of waste management studies for the dairy industry were carried out at different universities and factories throughout the country. One important observation seemed common. Most researchers were able to show that they could reduce wastage in plants by about half with little to no capital investment. What most of the studies did not report, however, was that factories were revisited and resampled in later years, and that waste reoccurred and people tended to slip back into bad habits. A common reason cited for waste reappearance was that the person who started the waste reduction program left the firm. The point is: It's important to follow up on new programs, and the role of waste management must be an ongoing full-time activity. A waste warden is an important job, and I think managers must recognize this.

It is no trick at all to reduce total waste flow volumes in most food plants by at least 50 percent. Some interesting findings occur when plant employees know water use is being quantified on a daily basis. In the Brooklyn plant, Figure 2.3 shows total water consumption was cut from 80,000 gallons to about 40,000 gallons. This would result in $40.00 savings per day, assuming water cost to be $1.00 per 1000 gallons. Monthly savings would average about $800.00.

To begin with, one has to know how much waste is being produced, where it comes from, and what its composition is. Knowing the "what" and "how much" frequently provides answers and opportunities for coupling unused resources into existing products.

There's little doubt in my mind that day-to-day wastage becomes so familiar in factories that we tend to think the events are normal and are unavoidable. It's not unusual to locate unseen, out-of-sight problems in a factory once managers initiate water use audits on a regular basis.

The Water Audit

I once read that "most leaks are easy to find, they can be so familiar that they eventually become like old friends!"

The expense for not taking the time to fix leaks will be initially reflected in your water, energy and disposal costs. There are, however, other kinds of leaks in some factories, the cost for which is not immediately appreciated because they usually exist underground, in walls, on roofs, and in other out-of-easy-sight locations. They probably occur from pipes damaged by corrosion, from frost heaving, or from worn-out valve packing.

What is important to you as managers is that these leaks be ferreted from the "out of sight, out of mind" waste areas because over a period of time such flow causes expensive structural damage to your buildings, parking lots, etc.

The best time to make an audit to find hidden leaks is when the plant is not operating. Water audits can be made on either the feed side or the discharge side of a water flow through the factory. If the plant meters its water (and most do), the job is easier. Readings at the start and end of a test time period will tell you the consumption amount. It's pretty straightforward from this point to account for operations that use water even when the plant is not in use, such as boilers or air conditioning.

If this option is not available, you certainly could look into sewers and manholes and see if water is running where none should or look at the outlet pipe or a main sewer line to measure how much liquid is being discharged per unit of time. You might also want to walk around the plant and listen at key points to find out if you can hear water running, even if you are not able to see any.

What has to be done to make a water audit follows the same kind of guidelines stressed in other waste conservation measures:

1. Draw a schematic layout of your water distribution system.
2. Show all points where water is used.
3. Measure how much water is being used at each location.
4. Tally water consumption per day over several days that bracket typical operations.
5. Compare the estimate of what you think is being consumed to what your water bill claims the usage to be.

Believe it or not, this methodical exercise, more often than not, picks out areas in the plant where controls used for regulating water are faulty and are not doing the job. One finds simple problems such as non-operating float balls in water tanks that are supposed to shut pumps off and do not, so water is being pumped continuously to overflow lines.

Sometimes water-accounting investigations point to unit operations that continue to draw water after the equipment is taken off the line. Processing vats equipped with zone heating and cooling coils are notorious for wasting water, usually because they're plumbed with so many valves that workers sometimes forget to shut the right valves off when switching from one heating or cooling medium to another.

At the top of our "watch" list, though, and perhaps the biggest water waste to be found in some plants, is the "overuse disease." If the water temperature on a discharge side of a cooling coil is about the same as it was for feeding the system, you can conclude you're using too much cooling water or the heat exchanger's surface is not working properly. Thermometers certainly ought to be installed in the piping to help regulate water flow. An overuse of only 10 gallons of water a minute is easy to waste with big line piping, and if only one machine does it for eight hours a day, you lose 4800 gallons. Like other diseases, the "overuse disease" appears to be contagious in factories if not treated. One medication is the water audit, and it ought to be repeated at regular intervals to verify that water is being used efficiently.

Statistical Process Control (SPC)

The same techniques designed to improve quality and efficiency in product manufacture can be used to manage waste treatment operations. In fact, we can think of the concept as an extension of the practice used by Dr. Demings in the early 1950s when he helped the Japanese to rebuild their industrial processes.

Unlike "run charts" that show daily variability in waste flows and concentration, or even more detailed information processes, control charts depict action steps. The work force need not be trained statisticians to use statistical control concepts. What employees given assignments to monitor wastage can do is

1. Collect data
2. Draw flow charts
3. Prepare histograms
4. Prepare an analysis chart that ranks data from highest to lowest ("Pareto Analysis")

The goals to be reached once data collections are complete would be to decrease flows and lessen flow variability. SPC charting techniques can be used to reduce waste loads on treatment systems, as well as to decrease episodes of discharge overloads that can disrupt proper treatment plant operation.

Like the control charts previously mentioned to control package-goods weights, a process control chart is meant to keep control of a specific unit operation.

The operator or supervisor is provided with a "run chart," such as Figure 7.6, on which upper and lower limits are established. When flow

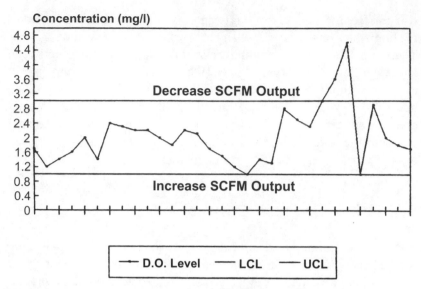

Figure 7.6. Daily run chart of dissolved oxygen readings where operator regulates aeration by controlling SCFM (air) output of blowers.

or strength indicators being monitored get out of hand the operator is advised to adjust the process. In some situations control limits are expected to range within three standard-deviation-from-mean values. This does not always have to be so. What is important to know is that some "expert" needs to establish the optimum range within which an operation can be run with a high degree of efficiency. Note the change in flow rate in Figure 7.7.

Of the many examples we can cite that illustrate value in using statistical process control systems, let me deal with some different industries.

In the case of a milk processing plant, two charts of flow rates are shown. The first, Figure 7.8, shows a plant prior to letting staff know what goes to the drain. Figure 7.9 shows the flow rate when the hourly worker is in control. If you overlay the two figures, you see that over 50 percent reduction and substantial savings were accomplished with almost no outlay of capital.

In the case of egg breaking operations, a process flow diagram like that in Figure 7.10 is shown; a five-plant work program developed by staff at Cornell University generated average improvements cited in Table 7.1. Water use dropped about 25 percent; weight egg loss per pound of liquid eggs processed decreased around 40 percent.

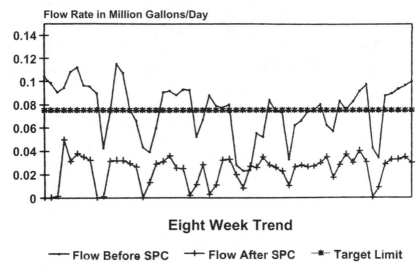

Figure 7.7. Flow profile of a food factory before and after a statistical process control system was initiated.

As in many industries, most of the plant managers did not know, nor did they have available methods to determine, the amount of waste generated in egg breaking operations. They did realize that their overall losses might amount to 6 to 10 percent of the total output, depending, as they thought, on the eggshell strength.

Even though egg breaking plants process less than 30 percent of the nation's eggs, and egg grading plants the remainder, losses from breaking plants exceed grading plant waste by more than tenfold, thereby presenting an equal or greater pollution potential.

Up to 15 percent of the total egg liquid output was lost to the sewer in plants where good in-plant management was not practiced. Losses equivalent to three eggs for every dozen broken were reported as maximum losses that occur in plants where no waste conservation measures were practiced. The average pre-modification product loss in all five plants sampled was 12.5 percent (by weight) of the processed output.

The average egg liquid loss in a medium-sized facility (two or three breakers) represents a decrease in revenue between $500.00 and $700.00 per day.

The losses on a product basis averaged as follows: Before in-plant waste conservation 0.034 pound BOD5/pound of egg liquid was produced and 0.90 gallon wastewater was generated per pound egg liquid

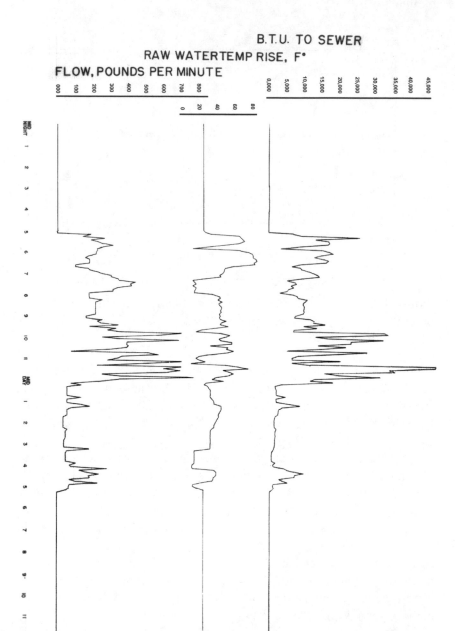

Figure 7.8. Flow, temperature, and BTU content of plant effluent as a function of time before water conservation practices were put in place. Employees were unaware plant volume was being monitored.

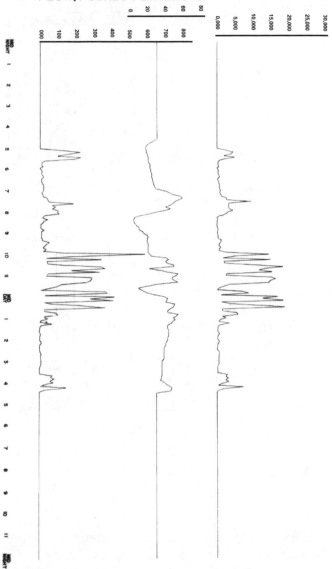

Figure 7.9. Flow, temperature, and BTU content of plant effluent after water conservation practices were put in place and employees knew flow was being monitored.

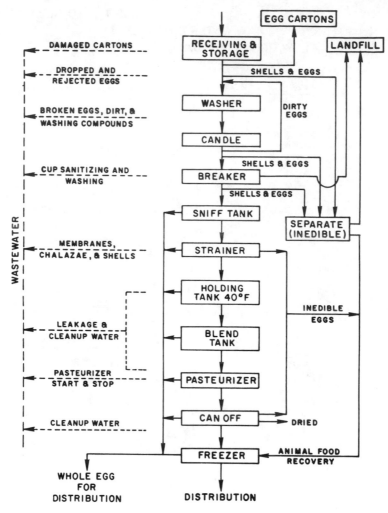

Figure 7.10. Flow diagram of product in an egg breaking facility.

produced. In-plant modifications decreased average BOD5 losses by 50 percent and decreased wastewater volume by 24 percent.

In-plant waste control was found to reduce the waste generated from an average of 12.5 percent product loss to 6.4 percent product loss. This is equivalent to additional egg product recovered worth between $250.00 and $500.00 per day in a medium-sized breaking facility, not including the savings from reduction in cost of waste treatment.

Table 7.1. Average unit production wastewater volume and organics generated in egg breaking facilities

	Plant A	Plant B	Plant C	Plant D	Plant E	Average
Before Modifications						
Volume						
wastewater (gal./doz)	0.536	0.884	0.959	1.23	—	0.90
Weight egg loss (BOD$_5$ [lb]/lb liquid eggs)[a]	0.0380	0.0345	0.0322	0.0316	—	0.0341
(lb [wt]/lb liquid eggs processed)[b]	0.14	0.126	0.118	0.116	—	0.125
After Modifications						
Volume						
wastewater (gal./doz)	0.689	0.646	0.279	—	1.13	0.686
Weight egg loss (BOD$_5$ [lb]/lb liquid eggs)[a]	0.0120	0.0228	0.0084	—	0.0243	0.0169
(lb [wt]/lb liquid eggs processed)[b]	0.044	0.083	0.031	—	0.100	0.064

[a]Calculated by assuming that the weight of liquid egg obtained from one dozen eggs was 1.21 lbs.
[b]Calculated as follows:

$$\frac{\text{lb BOD}_5 \text{ lost}}{\text{lb egg liquid processed}} \times \frac{\text{lb COD}}{0.58 \text{ lb BOD}} \times \frac{\text{lb egg liquid}}{0.46 \text{ lb COD}} = \frac{\text{egg liquid lost}}{\text{lb egg produced}}$$

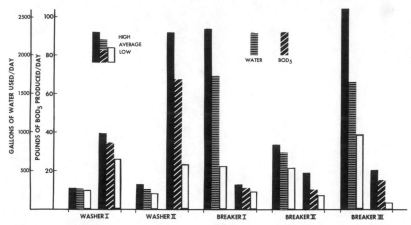

Figure 7.11. Mirrorlike egg processing equipment generated different waste loads from similar lots of eggs.

Good in-plant waste control would result in annual product recovery of 70 million pounds of liquid egg of a quality suitable for animal food which is now lost to the sewer, on a national basis.

The effectiveness of in-plant reduction of waste is not dependent on the size of the egg-breaking operation, since the portion of the product lost in small plants and very large plants was comparable, depending on the degree of management exercised.

To prove the point, Figure 7.11 shows that mirrorlike egg-processing equipment in a plant processing similar raw material (eggs from a common producer) generated different losses. The problem areas that needed adjustment were brush level and timing. Neither change would cost more than a few minutes' time to save substantial amounts of profit.

Figures 7.12 and 7.13 illustrate what was done in one factory and could well have been done in others.

Product Loss in Waste Steam

Color change in liquid wastes does not often indicate losses sent to the drain. Figure 7.14 shows that milk loss to water in varied amounts of dilution will fool you every time if you think color intensity is a measure of milk/water concentration. The case is not true for egg breaking plant wastes. In this case egg color does reflect loss concen-

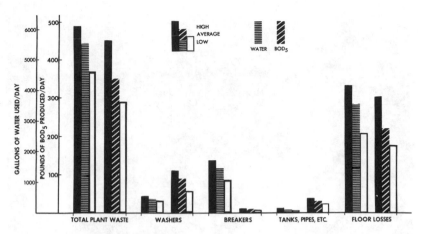

Figure 7.12. Volume and concentration of egg breaking wastes before a conservation program.

tration, as shown in Figure 7.15. Operators running egg-washing equipment would have done a better job if waste discharged from the washer ran through a section of glass pipe visible to the operator. Such a plumbing innovation might cost about 25 dollars per washer.

One quick way to check your factory's waste is by the color intensity check. If the loss is grape juice, prepare a dilution of juice in glass jars by having the volumes with water down to less than 1% and compare

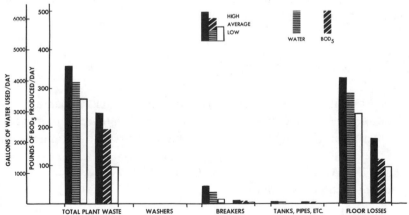

Figure 7.13. Volume and concentration of egg breaking wastes after a conservation program.

Figure 7.14. Ten bottles filled with different milk-water mixtures. Bottle on left contains 100 percent milk, while bottle on right contains 10 percent milk and 90 percent water. Bottles between are 90 percent milk, 80 percent, etc.

colors. The same method can be used with other commodities. If your effluent is color-dilution sensitive, you can have a quick monitoring check to keep an eye on wastage.

A Food Operation

Not too different is the case of an organization that prepared ready-to-eat specialty food entrees. The operation used statistical process control charts and came up with the following findings, actions and results.

STATISTICAL PROCESS CONTROL CHART
Findings, Actions, Results

FINDINGS:
Charting revealed high flows on weekends—

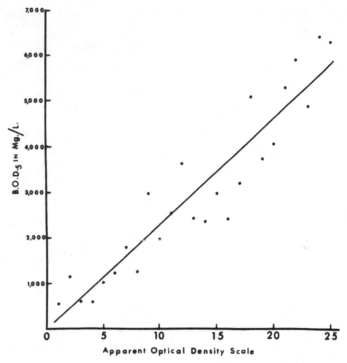

Figure 7.15. Egg color in waste effluent reflects loss concentration.

ACTION TAKEN: Preliminary Phase

1. Team assigned to investigate usage from plant process areas—
2. Monitoring and measurements on an hourly basis—
3. SPC run charts developed—
4. Set meeting with process areas to review run charts—
5. Developed loss notice procedures—
6. Implemented recommendations—

RESULTS OBTAINED:

1. Minimized water use on non-production days—
2. Reduced sanitation water usage—
3. Reduced influent flows to waste treatment—
4. Cost savings-operations/water/sewer user fees—

The SPC program was accepted for use in its waste treatment plant. Run charts, as well as operational charge charts regarding sludge volume index information and other parameters, were constructed. The goal was to assist operators as to how best to achieve treatment plant effectiveness.

RUN CHARTS:

- Each aeration basin control chart was developed to assist the operators in decision-making procedures.
- Upper and lower control limits were established, using historical data defining the best performance.
- Action procedures were developed for occurrences exceeding the upper or lower control limit.
- Operator variation (procedures) was reduced by having all operators perform the same function when control limits were exceeded.

Control Systems/Laboratory

Control systems include people and equipment. How do we justify control systems in the food factory?

First of all, the goal most commercial food producers have is to sell for profit those products they grow or process. It's a poor businessperson who doesn't make sure the product being vended is safe to eat, because sick or dead customers buy little or no products.

Just as process control systems are solidly entrenched in today's modern food plants to automate machinery to save time or money, we also have rapid chemical/biological test kits. These disposable tools can be used by laypeople to carry out quality control tasks to keep foods from spoiling.

Quality control instrumentation, which drives process control automation, can be networked with quality control (QC) programs of suppliers by a sharing of responsibilities. In fact, the idea is not different than the hazard analysis critical control point (HACCP) technique.

I've been told that some firms discard carloads of raw materials because the commodity being purchased didn't meet company standards. Two things come to mind. I assume the shipper was not paid for carloads, but where were the trainloads discarded? The act of throwing away food material is not cost-free.

For those managers running refrigerated foods operations know fully well the importance of "just in time manufacturing practices." This means that a food plant gears its operation to receive, process, and ship perishable finished goods on a daily basis. Very often, product shelf life can be as little as a few days.

These kinds of operations can let losses get out of hand unless rigorous quality assurance/quality control (QA/QC) practice is in place. Streamlined inspection systems (SIS) like those recommended by our USDA may be useful in dealing with meat and meat products. However, the fruit and vegetable segment is still better served by controls put in place at the farm gate. By establishing quality grades at the farm, we reduce wastage at the processing plant. Culls and substandard food stock are best left in the fields and not sent to food factories where they will become solid waste. Because we are now able to sort, cull, and grade foods on swiftly-moving conveyors using color video images, there's more reason to prevent off-standard raw material from being accepted.

We find that robots are now decorating cakes in the baking industry in order to produce foods with uniformity. In fact, we read almost daily about new improvement (so they say) in packaging, in food ingredients, in automation, and in process control. These changes are exciting and deserve to be considered for use in factories on an individual basis for many reasons, the least of which include automated controls for better waste control strategy.

Tests That Count

For some firms, the investment in staffing QC laboratories with chemists and microbiologists was thought to be an excessive overhead expense. We know, for example, canning plants get by with few laboratory diagnostic tests other than hot shelf incubation tests. Lots showing visible can defects are suspect and segregated. Unless canning operations are involved in research and development studies, laboratory facilities are usually small and staffed without formally-trained people. Of course, the firm can bring in additional part-time labor when required.

The other side of the testing picture might be QC methods demanded by the dairy industry. The law dictates that a full range of chemical and biological analysis be carried on raw milk and pasteurized products. Thus even small milk companies provide extensive laboratory facilities versus operation size. Sometimes QC labs that monitor food are used to monitor waste treatment facilities, if the firm operates its own system.

Why Me?

Managers are often dismayed when they compare QC practices of their operation against food plants of others. The key is "the law doesn't require that a product has to taste good, but rather it has to be safe to eat." This being the case, if you produce a food that is thermally protected against disease-causing problems, such as in canning, all you have to monitor are time/temperature processes to get the correct "D" value in retorts.

Unfortunately, when you vend refrigerated perishable food you now need to be concerned with a myriad of environmental contaminates that use the food as growth substrate, and the end product can be troublesome.

Up to now, we have been skirting the issue of waste management in the discussion of the QA/QC function. Just as test kits were developed for food process monitoring systems to minimize technician time to promote laboratory efficiency, so are the same types of kits available for water and wastewater testing.

Food plant managers can provide the waste treatment supervisor with laboratory instruments that are portable and disposable. A full array of new chemical and biological testing products can be used as tools that take away "black box" secrets from the biological process of treating waste.

If a firm has more than one factory, it can use a mobile laboratory to service the needs of its satellite operations. European companies and firms in the down under countries (New Zealand and Australia) have used the idea with success. In the past only big organizations, like Kraft Foods or Unilever, set up in-house waste management divisions to oversee multiplant company practices. This is no longer true because even smaller organizations need to be able to take care of their waste management practices. Sometimes the smaller firms use outside consultants to help them take care of environmental constraints, but more firms now elect to use in-house staff to manage waste treatment facilities. This being the case, it makes sense to take advantage of prepackaged test kits to run the chemical/biological analysis needed to control and monitor waste streams.

CHAPTER 8

Converting Costs into Credits

One Success Story

Waste streams from one food processing operation can become the feed product for another. Milk and its by-products are the classic examples of this practice. In the last 25 years we have seen a technical revolution take place in the dairy industry, enabling the division of milk into many different useful and beneficial products. Milk, cream, butter, cheese, and powder were the main products of commerce until more recent times, with each commodity wasting fractions of its product.

This author believes that the very fabric of the separation, fractionation, and concentration system changed with the use of membrane systems. Traditional cheese-making practices were shattered when "cheese-making artists" accepted the idea that milk and milk components could be concentrated at almost room temperature, and made into everyday cheese products with lower cost, better yield, and less waste. In fact, membranes could be used to alter milk composition in such a way that less whey would be produced when milk was converted to cheese.

Whey, as a waste, could be further fractionated so as to skim off whey proteins, which were worth many dollars to the industry.

Deproteinized whey can be, and is, used for feedstock to produce lactose. Other uses for the feed material are conversion into ethanol, acetaldehyde, methanol, and a myriad of organic acids. Biotechnologists use whey by-products for substrate media in fermentation systems. Other than single-cell protein, the list is large and the rewards have

been substantial. I'm pleased to have been able to spend some rewarding years as one of the participants and as developer of a process to
convert dairy waste into beneficial credits.

Separation systems, where possible, can be an effective way for food
plant managers to start exploiting waste expense into credit-producing
activities.

Common sense dictates that we begin taking hold of the waste problem by producing less. Sometimes the easiest way to reduce waste is to
be sure feed impurities are not allowed into a process without some purification step. Soil residues as part of crops, for example, need to be
left in fields and not be carried into food operations.

When possible, managers should analyze wastage to decide if waste
streams can be recycled back to process as part of a feed stream. Sometimes recycling steps will increase process efficacy, thereby increasing
yields.

The least any manager ought to be doing with waste flows is to know
waste composition. Waste streams can probably be fractionated into select components, some of which may become useful by-products. Even
when separation spin-offs seem to have no immediate purpose, the
process can create effluents that would be easier and cheaper to treat.

Reuse Systems

Waste characteristics between food factories differ depending on
product and process. What might be common is that food-handling machinery needs to be cleaned daily with detergents and sanitizers. While
we see more and more organizations using clean-in-place systems that
recycle some of the cleaning chemical, few firms go to the trouble of
segregating spent cleaning material for special purposes.

Acid/Alkali/Neutral

Cleaning liquids are almost always alkaline or acidic with few exceptions when neutral material is used. Because soil materials would be
organic or inorganic, the industry elects to use caustic and acids to assist cleaning protocols. It is not uncommon to see 2–5 percent caustic
solutions prepared from bulk 50 percent, a caustic supply for washing
multistory evaporation systems. The acid rinse to remove mineral de-

posits might be phosphoric or even hydrochloric formulations. Here too, bulk acids are used to prepare cleaning liquids.

Some cleaning material is purchased in drums or bags as ready-to-use material. As a rule, these chemicals contain water treatment compounds and are used in low-strength amounts.

The economics of using recycling cleaning systems are not often clearly defined. Initially, factory managers switched from manual cleaning systems to Clean in Place (CIP) to save labor. What was not apparent to early users of the change was that the factory would use manyfold times the amount of cold and hot water as well as much more of its cleaning chemicals and that capital funds were needed to automate CIP systems. Nevertheless, even with all the extra costs, I doubt people would revert to manual cleaning. What was not put up front with CIP systems was that a lot more water (hot & cold) and cleaning chemicals are sent to floor drains that need waste treatment. In other words, the increased hydraulic load to treatment plants can be mostly due to CIP discharges. Factory managers do not often identify these dollars. It is also troublesome that CIP circuits flush the same amount of liquid through cleaning operations regardless of the amount of soil left on equipment. The cleaning stages, which are rinse, wash, rinse, and sanitizing steps, etc., are the same unless a system is reprogrammed. Most systems adopt overcleaning steps because time is important and the factory needs to be back on line for processing foods after the typical 4-hour wash-up time allotment. The manager prefers to overclean rather than risk inadequate wash-up.

Cleaning Technology Has Not Kept Pace

No issue of proper low-cost cleaning has been more abused than that of washing membranes used in food factories.

The industry is dependent on a "black box" cleaning technology now being carefully orchestrated by only a few players. On one side of the issue is the supplier who wants above all to place a reliable system in the plant to clean in a way that promotes membrane life. The goal is noble, but is it efficient? On the other side we find users or buyers who need to question just how much cleaning expense they can really afford to put into the costs of operation to wash membrane systems.

When equipment is offline, it is lost time and, as such, adds substantially to operating costs. Of no small importance is the fact that membranes do degrade over time, regardless of use. Isn't it better in the long run to keep them operating?

We need to raise the question as to how much time it really takes to clean membrane plants. Some technicians and engineers seem happy to work out cleaning programs that gush and flush membrane systems with lots of treated or conditioned water and costly chemicals and then go on to squander energy with very long wash cycles.

We can waste disproportionate blocks of time and energy by pumping cleaning solutions around in piping longer than necessary once the soil breakaway phase occurs.

Only marginal benefits occur after some optimum point has been reached, regardless of motor size. Bigger is not always better.

Most costly of all items, although not normally identified, can be the "opportunity cost" incurred by downtime and not running with revenue-producing products.

Ideal Cleaning Methods

Cleaning protocol varies somewhat with site conditions and food commodity, as well as with detergents and sanitizers, but procedures should not be so different that cleaning the system appears mysterious.

I recently surveyed an organization in Europe that operated five membrane units in different countries including the United Kingdom, as well as sites on the continent. Much to my surprise, and theirs too, we came to the realization that cleaning methods in all plants were markedly dissimilar.

Cleaning methods differed as to types of chemicals, time/temperature points used in the cleaning cycles, and expectations of what was considered to be a satisfactory cleanup job by local managers.

The truth of the matter was that no attempt whatsoever had been made by people in the organization to standardize equipment cleanup methods even though three of the five plants were processing the same commodity, sweet whey.

Chemicals

In some cases we can probably adequately clean membranes using only caustic and chlorine solutions. From time to time some organic acid might have to be flushed through the membranes to take care of flux decline rates. This does not mean that we might not need, on occasion, to incorporate wetting or chelating agents into the different cleaning compounds.

Obviously, there can be advantages and real value in working with some firms that seem to be specializing in formulating cleaning supplies for use in membrane equipment.

What I'm not so sure about, however, is whether enzyme cleaners once used to clean cellulose acetate systems are now needed. From time to time enzymes can be put into play, but only when they are really required.

Importance of Cleaning Systems

When cleaning systems are left to drift they can complicate wastage and will be more costly than is being recognized by people operating food-processing operations. There's no reason why cleaning chemicals cannot be reused, especially when only about 10 percent of the cleaning capacity of a solution is dissipated by a normal washing cycle. The trick is to create an environment to separate food residues picked up in cleaning loops. Debris from cleaning can be removed by screening, settling, centrifuging, etc.

Figures 8.1 and 8.2 illustrate dissipation of an alkaline cleaner when recycled without addition of added caustic after each run. Microbial increase in cleaning solution is also indicated.

In practice, the CIP system is charged with sufficient cleaning chemicals to accomplish the cleaning function. A biocidal chemical concentration is selected. After the solution is used only additional cleaning material is added to keep the solution at its predetermined concentration. Debris removal that cleans the detergent liquid will lengthen the recycling period the solution can be used.

Spent Cleaning Liquids

Food plants that need to pH-adjust their waste effluent prior to discharge to municipal or private treatment plants should use waste acids and alkali for this operation. Sometimes these liquids are too dilute to be useful; but, by capturing only first flush, rinsing makes it possible to save only the more concentrated material. Cleaning chemicals, like water, are

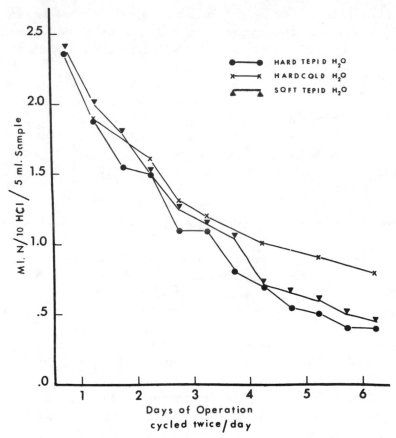

Figure 8.1. Apparent detergent dissipation of an alkaline cleaner when recycled without being recharged with additional cleaning material. Hard and cold water were used.

a two-way sword. There is an initial cost to buy cleaners, and treating detergents costs the firm money. Detergents in themselves contribute BOD loading to the effluents.

Other Cost-for-Credit Opportunities

If we exhaust ingredient salvage considerations, do we stop looking for other techniques to become more efficient? Most food factories probably use steam, refrigeration, and electricity which, along with other factory services, need to be evaluated for cost-effectiveness.

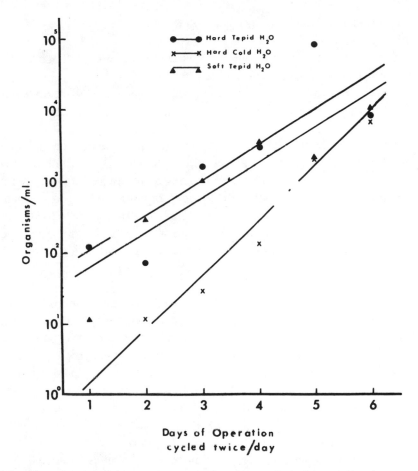

Figure 8.2. Microbial contamination of recycled detergent when not recharged with additional cleaner. Test made by standard plate count. Hard and cold water was used.

Energy Conservation via Waste Management

Just as supervisors need to address a series of questions regarding product waste, they need to look at energy usage. There's a real need to think in terms of "fuel for thought" because there are methods to economize variable-cost items. Energy conservation begins with a use inventory.

1. Determine electrical energy usage at site.
   a. Learn how to read and understand billing.
   b. Find out where energy is being used.

c. Evaluate how changes in implant flow schemes might decrease power consumption.
d. When was the last time someone looked at energy use? Is this an annual exercise?
e. When did you last invite power company personnel to the factory to review your electricity usage?

These and other thoughts need to be put on the table in open dialogue with you, plant engineers, and the utility company. I can think of reasons to own substations with transformers and reasons not to own them. In some cases it may be to the advantage of organizations to consolidate power use by running poles and power lines to off-site waste treatment plants. The main benefit to the organization is to get full input by line supervisors into the loop to think about energy consideration.

2. Determine Fuel Consumption:
   a. Supervisors ought to know what kinds of fuel the factory utilizes. Common knowledge by supervisory staff should be that they know how much energy each type of fuel provides. This information needs to be defined in BTUs or pounds of steam assuming 100 percent efficiency.
   b. The factory ought to have steam flow meters to quantify power production. These data are tabulated at least monthly and compared with fuel consumption to determine usage efficiency. We must assume the plant engineer is able to measure boiler efficiency at the combustion box.
   c. Some factories find it useful to establish load centers within factories. I installed flow meters at key locations in milk factories I managed over 40 years ago when bunker C oil cost 5+ cents per gallon, and found the metering scheme economically worthwhile.
   d. Ask the question of whether or not steam usage would be reduced by in-plant improvements—that is, longer process runs with improved scheduling.
   e. Determine whether or not steam or hot water leaks exist in piping or plumbing. The same loss information needs to be done with equipment that uses steam or hot water.
   f. Determine energy use/unit of time for machinery and compare actual use versus theoretical.
   g. Do you have energy consumption standards?

Both electrical and steam use data can be gathered by individuals with very basic equipment.

By using an amp meter, one measures machine amperage load and multiplies amps times voltage, which yields watts. Kilowatts used times price/kilowatt equals cost. Any factory can generate this kind of information.

Even without flow meters, we ought to be able to collect steam condensate and weigh it per unit of time. This kind of information will tell you the pounds of steam a unit is consuming to process some amount of product. Pounds of steam times cost per pound of steam yields heating expense expended to process.

The purpose of each of the exercises cited is to let the worker know whether or not plant efficiency is peaking or lessening.

CHAPTER 9

Economics of Managing Food-Processing Waste

Pick up almost any report from the proceedings of a food waste conference and you will probably read, "The most effective waste management techniques are source reduction." These themes repeat themselves over and over throughout this book. No waste treatment system will ever be better than a system put in place to stop waste.

Production workers, shift supervisors, midmanagement, and others need to become aware that waste and its spin-off costs represent loss profits to the organization.

A Double-Edged Sword

If we begin with the premise that food originates from agriculture, there's a raw product cost for cleaning and making these commodities both aesthetically and sanitarily safe. To begin with, agricultural products require large amounts of water to initiate a first step in a processing sequence to process food. In much of the region where farmers raise crops, we should note that irrigation uses enormous amounts of water, but we will not make this an issue in this text.

Process water in the food factory is really a double-cost utility because it costs money to get water and then more money to get rid of it as waste. We might estimate dollar water value to a food firm to range from 40 cents to two dollars/1000 gallons. The cost to deal with wastewater is about the same as raw water except when pollution loads are high and surcharge costs are tacked onto service charges.

Firms that need to pump water from their private supplies—like lakes, streams, or wells—experience pumping costs. If the firm treats its own sewage, the costs range from a few cents/1000 gallons to several dollars depending on geography, state ordinances, etc.

I'm not sure we can get the attention of factory workers and shift foremen with a detailed cost analysis of dealing with waste. Unless the employee was on hand when a factory was built from the ground up, he or she should not be expected to visualize an empty tract of land without a building.

Sunk and Variable Costs

Capital costs of land, building, and equipment are for the most part "sunk costs." When I took economics in college back in the "Dark Ages," the professor stressed, "sunk costs are not used for decision-making purposes." Years later I pretty much agree with this statement. What is tangible to the work force is what we call variable costs, which accrue as a function of plant operations. These costs are power and other utilities such as steam, air, water, etc. We could add to the list chemicals to treat waste. Substantial value probably could be assigned to food fragment losses from peeling and cutting operations, were we discussing fruits and vegetables. The loss might be whey if we were manufacturing cheese. Perhaps we ought to note the loss from cooking processes of specialty food products. These food items do have unit value per pound, and they can be assigned value based on raw material cost and finished goods yield per pound of raw product processed.

Okay, why not start with an analysis that claims monetary value for product wastage based upon some percentage recovery value.

Operating Costs

Additional costs that need to be added to waste components are the actual operating costs. Operating costs include labor, power, repairs, taxes, and other expenses. These expenditures add up to the cost of doing business, as well as the sum of the total cost to manufacture a product. We could argue whether the loss is valued at sales value, wholesale, or actual cost amount. Some authors suggest the cost of product components should be calculated on the basis of prevailing market prices at time of analysis.

Sharing Information

We cannot criticize the fact that managers are not willing to tell workers what the profit margin may be for foods being manufactured. In fact, would you want to tell your work force or competitor what it costs to make product X? The waste value can be created with some validity by broad-brush accounting methods. I would think this kind of information could be shared with the work force without giving away family secrets.

We do ourselves a disservice if we let the factory work force think potato scraps, for example, are only worth a penny or two a pound when the waste actually costs the firm much more than its value at the farm gate. Workers understand transportation expense, processing costs, cost of capital, etc. What we need to do is to teach by show and tell methods that plant activities cost money, and what we fail to do properly costs the operation money. An informed work force is a better asset than personnel who fail to appreciate real value of the production.

Treatment Economics

How many people in your organization know what it costs the firm to deal with its liquid and solid wastes? Have you lumped these costs into general overhead expenses, or should you credit these dollars the value of lost components? For organizations lucky enough to be able to use publicly owned treatment works (POTW), the expense is readily available by evaluating the monthly or quarterly billing against waste loss figures. In other words, you have a waste cost index of real dollars.

Some firms elect to use pretreatment systems to reduce their municipal waste treatment costs. The goals might be pH adjustment, temperature control, solid screening, or aeration/sludge separation to decrease BOD amounts. There are capital and variable costs to these operations, which add to waste cost index. At this point wastage includes private and public treatment expenditures.

Should the corporation need to operate full-scale treatment systems, then these costs vary as to type.

The cost to construct waste treatment systems in new factories for the food industry in years past is 10–20 percent of the total amount invested in buildings and equipment. Annual operating and maintenance costs are substantial, costing many thousands of dollars per year. Waste volume and type of treatment system used to treat food plant wastes affect costs, as does plant location. A plant with a rural location may be able to operate at lower costs than city plants, but this is not always the case.

Information provided in Table 9.1 describes treatment type, effluent volume and cents costs per 1000 gallons. The annual costs for food industry factories show (regardless of unit price at the time data were collected) that when waste can be land-applied, the costs would be similar at the four levels shown. Treatment plant size and flow-through volume affect cost that is less per 1000 gallons with bigger operations. Plants fortunate to locate in major cities should be able to enjoy lower treatment costs than those trying to build and operate private facilities. The choice may not be there in the years ahead as municipal plants become outdated and new works need to be built. Federal grants awarded to municipal systems pretty much carry the demand that each user pay its fair share of operating the location. Fair share means percentage waste load of the overall municipal system.

Sewage Surcharge

Local governments are putting a new twist into computing sewage charges by adding a tack-on rate, which they call a surcharge to allocate extra treatment costs to sewage found to be stronger than their domestic allowed strengths.

Table 9.1 Annual costs (Cents/1000 gallons of effluent based on effluent concentration of 2000 mg/1 BOD)

	Volume of waste equivalent processed by plant (GPD)			
Type of Treatment	50,000	100,000	250,000	500,000
Activated sludge	351	198	114	81
Trickling filter	288	177	101	75
Aerated lagoon	175	101	57	42
Spray irrigation	140	140	140	140
Ridge and furrow	35	35	35	35

The way surcharges are calculated in your area will probably be spelled out in some detail in the local sewer-use ordinance. I suggest that someone in the firm be delegated to review the document at regular intervals. Strange as it might seem, even the big "customers" of our public treatment plants fail to attend city council meetings to take part in local legislation processes when surcharge fees are being established. As a rule, a notice of meetings of hearings to consider changes in service rates is published in local newspapers. Since we use large amounts of water and pay much for treating wastes, it seems reasonable to urge management to put their houses in order to reduce waste and expense. In addition, we have to ask city authorities the right kind of questions about how they go about establishing water and sewage rates.

Pyrolysis

How about using food processing waste to produce fuels like methane or petrochemicals to power factory boilers? This is a reasonable question often posed by supervisors. Why not burn solid waste generated in the factory, such as cardboard, wooden pallets, spent packing, etc.? I suppose the answer to these questions can be prefaced with some "what if" statements.

Let's look at the easy choice first. What if we decide to burn combustible wastes?

(a) Can we use the heat for processing? For heating water? For heating the factory in cold weather?

(b) Will we need to build an incinerator? What will it cost to construct and maintain a modern incinerator? What approvals are needed from EPA to prevent air pollution? What will we do with ashes and how much ash will the waste produce? What will it cost to burn waste?

(c) Will costs versus return on investment favor the burning of wastes?

I suspect if a positive answer from item c is large enough to be worth the effort, a firm might decide to take the plunge. The whole issue of solid waste disposal, recycling emphasis, and community action to convert wastes to beneficial purposes dictates action. Landfill costs are rising ever higher, so much so that burning may be a preferred alternative.

While we have not looked in sufficient depth at waste combustion by incinerators, my opinion for this route is one that will be more attractive in the years ahead. We should remember that many municipalities burned wastes prior to World War II both in the States and abroad. Incinerator ash amounted to about 30 percent of the weight burned. The ash needed disposal and was mostly buried. Ash disposal is no small problem, but less troublesome than dealing with the total mass.

Gas Production

As for the idea of producing methane and petrochemicals, there's an interesting comparison of energy use in aerobic and anaerobic processes to deal with sludge.

Figure 9.1 shows aerobic digestion as energy intensive, while the anaerobic methane produces energy, which then can be harvested for use in combustion operations. The idea of using sewage gas as a fuel is thousands of years old. Both the Chinese and people in India piped methane gas from dung piles with wood pipes for cooking long before they knew the biological pathways to produce the gas. The formation of methane gas, or marsh gas, has been occurring in nature naturally for many thousands of years. Since the oil crunch of the 1970s, interest in methane has been put on the front burner.

As far as waste treatment, for the engineer, the idea of stabilizing sludge and production of methane gas was standard operating practice. The fear of not having fuel in the 1970s, as well as the skyrocketing of energy costs, motivated food plant managers to think about converting carbohydrates, lipids, and other waste fractions into a gas supply.

Complex organic compounds can be converted to methane and carbon dioxide gas streams in about a 50–50 split with a modest capital outlay. If the firm wants to compress the gas, then a major part of the energy produced is used to compress methane; thus, it would be attractive to use the methane gas as produced.

Over the last few years, we have witnessed the creation of operations that produced methane to power boilers to produce steam. Methane is used to power generators to produce electricity. The conversion is not trouble free, because impurities in the waste and gas tend to be corrosive. Special engineering solutions need to be built into the equipment processing the gas.

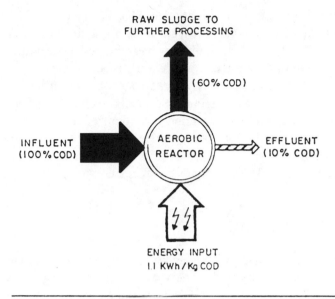

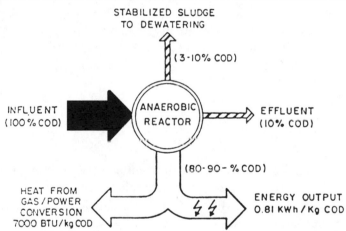

Figure 9.1. Energy use in aerobic and anaerobic processes.

Anaerobic Systems

Even though engineers recognize the value of anaerobic systems, and food industry effluents are amenable to such treatment, we have to wonder why more installations are not in place.

The reason is mostly that poor experience has been the rule for operations with anaerobic systems. They are more difficult to keep on-line (prevent from going sour). Bad publicity has hindered more widespread use of anaerobic waste treatment plants in the food industry.

Being unable to run methane operations with ease probably was caused by lack of information about the special biological requirements. Until recently, for example, few engineers recognized that minute traces of molybdenum (parts/billion) were needed by some anaerobic species to survive. In other words, for some species, special nutritional components are needed for fastidious organisms in anaerobic systems.

Hardware

Over the past 30 years, great advances have taken place in anaerobic treatment of food-industry wastewaters. Most notable changes were the configuration changes made to anaerobic reactors, which decreased the time it takes to put methane-producing plants on-line. Technology has been developed that increases organism numbers to substrate food so much so that with heat, degradation is accelerated from days to a few hours. Flow-through reactors are built to accommodate high strength and low strength (dilute) wastes.

Gas Collection

A recent innovation that seems to be arousing interest is to modify traditional anaerobic lagoons by covering them with butyl rubber-supported covers and collect the methane gas. Operations are on-line in New Zealand and Australia.

Figure 9.2 shows a photograph of lagoons being modified in Werribee, Australia. This is a world-famous sewage farm that treats waste from the 3.5 million people in Melbourne. I visited the operation in the spring of 1992.

The covered anaerobic lagoon serves two main purposes: (1) covered anaerobic lagoon smells less; (2) gases can be collected from anaerobic systems to be burned to rid the system of the gas or to tap the gas for beneficial purposes. The technology deserves our attention.

What may turn out to be a viable system for food factories is one that uses large land areas to treat wastes. My mention of the Werribee,

Figure 9.2. Lagoon being covered with a rubber material to trap sewer gas so methane can be burned.

Victoria, sewage farm in Australia is a classic case that's been on-line since about 1900. Other sewage farms were recorded outside of London, England, over 300 years ago. More recent interest in the idea has been expressed in our own east coast states. One organization proposed an "energy farm" system where industrial wastes would be blended with municipal waste, and different unit operations would treat waste at a profit. Profit would be realized by converting waste to gas to drive generators to produce electricity. The principals involved in the proposal have patents on new type-efficient reactors to convert waste organics into fuel. At the time of this book the plan is still under review.

Composting

Look around at your neighbors and you're apt to see composting practices that convert yard waste (leaves, grass, etc.) to a humus-like substance. Nature carries out the process naturally, and it's all around us.

About 75 years ago full-scale industrial composting plants were put into use in Europe. Holland and Denmark built closed composting systems for the reclamation of organic wastes. The Dutch formed a utility company and generated power from their operations. Some claim that, when composting is properly done, nuisance-free systems can be used to recycle organic food wastes into soil-quality material. The preferred method seems to be that which mixes organics in garbage with farm or garden wastes and then mixes the mass often enough to allow air to penetrate piles of waste stock. The process is exothermic, which pretty much controls the potential danger of disease as contaminants. The by-product is earthy odorless mulch that can be used for soil improvement purposes.

As with all biological properties, there are basic principles to follow that attract composting microorganisms. The process requires carbon and nitrogen with about a 20:1 C to N ratio. Often, peat and wood chips are mixed into the mixture to be composted. The method is probably best suited for rural areas where some food factories are located. There needs to be room to construct aeration piles spaced far enough apart to allow use of a tractor to turn piles over, so as to economize on labor. Success stories appear often in trade journals of operations that look favorably to the technique. More recent attention is given to fin fish processing waste where the use of composting is advocated as economical, effective, and environmentally safe. If fish wastes (an odorous commodity) can be processed by composting, I would think more food plant solid waste could be handled using similar operations. What is important for a firm to remember when dealing with composting is that composting is a microbiological process that requires bacteria, fungi, and actinomycetes to make the process work. Factors like moisture, oxygen, pH, temperature, and carbon/nitrogen ratio keep microbial populations working.

Just as a waste treatment plant operator needs to follow rules to keep the unit operation efficient, the same is true when wastes are degraded using composting. The system needs to be managed with appropriate resources of space, labor and substrate.

CHAPTER 10

Training

Employees, including the hourly worker, supervisors, and managers at all levels, need development training on an ongoing assignment to keep firms viable to compete at home and abroad.

Some organizations identify management training and development as critical success factors. Human resources within an organization are recognized as assets and are nurtured accordingly. Major corporations such as Kraft Foods International, for example, create departments of human resources and employ directors of the departments to maximize human resources. It's no secret that large amounts of capital are used annually for training and development of managers. We can question whether or not sufficient funds and or energy are invested into line or blue-collar staff within these same organizations. *Newsweek* wrote in its Sept. 21, 1992, issue that "most of the money U.S. companies spend on training is spent for executives and managers, not for production workers," says Rutgers University Manpower expert Carl Vanhorn. Firms engaged in international trade (and which firms are not in these days of global marketing, when products are fabricated with a myriad of components not produced on-site?) are continually claiming a need to become "leaner and meaner" to compete abroad.

There's a consensus among people who track employee training activities that firms with a definite commitment in ongoing training programs for their managers and non-managerial employees are more apt to be successful organizations. On-the-job training is not one short exercise, but needs to be looked at as an activity providing employees additional training throughout the individuals' careers.

Pride Is Important

Training workers to develop some pride in their occupation in the food field is a real task and concern for management. A firm, providing an ideal plant with excellent equipment, will do itself a disservice if it selects persons for its staff without thought to their interest or ability in working with foods. A single, competent manager is not the cure-all. The manager who believes he or she alone runs the plant or the section of the factory will soon find that work is accomplished through others simply because the manager cannot do all the jobs around the clock, 7 days per week. As for the worker: age may not be the factor to program training. For my part I see people that can be young at 60 and old at 40 in the way the individuals are willing to adapt to new ideas, technology, and philosophy.

As for our young first entry worker, it's probably true the new worker is fast, quick to learn, and generally mechanically inclined. The new young employee has been a television watcher from childhood and, having been conditioned to computers, robots, and the like, can hardly be expected to become enthusiastic about jobs like scrubbing walls and floors, no matter how vital it is to the operation. Given a job opportunity in a more alluring industry, the individual will soon leave. No wonder people in some companies were upset when I was writing monthly waste management articles in a trade journal and I suggested we put educational information in employee lunch rooms, as well as in the lobby for visitors. Some readers wrote to me that they didn't think it was a good idea. The complaint was that the factory worker was more apt to look at employment ads and salaries and then either leave or ask for increases in wages. The truth is obvious, and it's a given fact that we need to address the issues close to home in the food business if we expect to retain and keep the best talent of the work force.

Some years ago (1979) three extension specialists (Professors Carawan, Chambers, and Zall) at three different universities (Cornell, Purdue, and North Carolina State) collaborated in putting together a series of fifteen manuals dealing with water and wastewater management subjects. Their goals were to teach food processors, engineers, scientists, waste management specialists and other practitioners the concepts and principles needed to control waste in food-processing facilities. The program was funded by the Science and Education Administration Extension USDA. The work generated fifteen modules with a core unit prepared to introduce the program. Technical specifics were presented in seven modules with commodity treatment in seven additional units. Two sets each were delivered to extension agencies in each of the 50 states so they could be used for providing training materials to food-industry operations of the different states. The three professionals put short courses, as well as national training schemes, into place. While there were benefits from the program, it's fair to think the idea of waste management for factory managers was not fully embraced.

Professor Carawan, for example, developed a training program for food plant managers to help them control water and waste. The efforts centered on a 10-hour training exercise which dealt with a management phase, a water supervisor phase, an employee phase, and a follow-up phase.

An outline of the program is listed, but its contents can be tailored to fit specific site problem areas.

Water And Waste Management-10-hour Training Program

MANAGEMENT PHASE	– 2 Hours
WATER-WASTE SUPERVISOR PHASE	– 2 Hours
EMPLOYEE PHASE	– 4 Hours
FOLLOW-UP PHASE	– 2 Hours

Four Reasons Why a Food Plant Needs a Water And Waste Management Program:

1. Water Conservation And Waste Reductions Are Important Economically
2. Management Is Often Unaware Of A Problem
3. Management Lacks Experience In Solving Water Waste Problems
4. Most Excess Use Of Water And Loss Of Product Is Due To Carelessness Or Uninformed Employees

MANAGEMENT PHASE

The specific steps to be taken are as follows:

1. Recognize the need for water and wastewater management effort emphasizing cost and product losses.
2. Examine water and waste terminology—bod, cod, tss, surcharges, etc.
3. Explain key governmental regulations.
4. Explain wastewater management and how to start a wastewater program.
5. Define the water-waste supervisor's responsibilities and training.

6. Emphasize the importance of employee education in successful management.
7. Apply cost benefit concepts to wastewater management.
8. Expect plant to reduce water usage and water bills by 50%.
9. Expect plant to reduce its waste loads and related surcharge bills by more than 50%.
10. Expect improved public relations, as plant becomes a better corporate citizen.
11. Expect employees to be more knowledgeable about plant operation as a result of greater emphasis on water conservation and waste reduction.
12. Expect improved plant profits as product losses are reduced and as water and sewer charges decline.

WATER-WASTE SUPERVISOR PHASE
Details are as follows:

1. Review of information presented in management phase.
2. Outline water-waste supervisor's responsibilities.
3. Explain the necessity for adequate record-keeping.
4. Point out key ways to keep management informed.
5. Emphasize the need for employee motivation on a continuing basis.

Responsibilities for the supervisor are as follows:

1. The plant must be surveyed. Sketches, drawings, and maps should be detailed that indicate the size, capacity, and location of water lines, meters, sewer lines, junctions, manholes, and other parts of the system. Operating data should be compiled relating to water used, water generated, and the bod and flow from the main sewer.
2. The water-waste supervisor and a management team should examine the plant critically. Questions, which should be answered, include: where is water used? Why is water used? How much is used? Where does waste originate? Why does waste occur? One effective method of evaluating a plant is to study photographs of the normal daily operation.
3. The management team should formulate a plan of attack on water and waste problem areas. Priorities should be assigned so that the most important projects are undertaken first.
4. Simultaneously, a water-waste educational program should be instituted. Employees must be informed and involved in the program. The supervisor must be responsible for the program but it is the operating personnel who will determine if the program will be a success.
5. If pretreatment or treatment is required, the water-waste supervisor should recommend that an engineering firm be retained to suggest alternatives.

6. The water-waste supervisor must exert a continuing influence on the plant operations and personnel.

EMPLOYEE PHASE
It may involve all or selected employees.

1. Illustrate poor practices, utilizing photographic slides of poor processing techniques to focus attention on problems.
2. Explain the need for water conservation and waste prevention, and emphasize benefits to employee and community if attack on waste-water problems is successful.
3. Encourage management to express its interest and concern for water conservation and waste prevention.
4. Define wastewater terminology.
5. Involve employees in the attack on water-waste problems by encouraging recognition and solution of plant problems.

FOLLOW-UP PHASE
This is a joint session with management and the water-waste supervisor.

1. Review program and evaluate progress in plant.
2. Tour plant for signs of progress and/or evidence of new or continued problems.

Poor Work Standards

Probably no food firm exists that is not troubled daily with operational problems created by its own staff of men and women. There is a shortage of trained technical persons, which affects the economics of production and sales. Trained workers are needed who are unafraid of horns, buttons, and lights and who have professional pride in their specialty. Conditions are so bad in many plants throughout the country and in other areas of the globe that even mediocre technicians have gained positions of prominence. It is not acceptable that top management needs to compromise on questions of poor work quality. Poor work quality, more often than not, is not the fault of the new or old worker. Our people in on-line positions of floor personnel expect training from competent, articulate supervisors.

Training, for factory managers as well as others in the operation, means the individuals are expected to acquire a myriad of skills appropriate to the organization's requirements. The old adage "if the student

hasn't learned, the teacher hasn't taught" is still true. If we allow training to be hit or miss duties then we can expect poor return on time
spent. When a new worker is told "go with Jim or Mary; they will show
you the ropes" more often than not bad habits permeating a work force
will be transferred onto the new worker.

First of all, every position in a factory needs a job description. The
worker assigned to a position has a right to know what is expected from
him or her. What are the "minimum" responsibilities the assignment
entails? Job descriptions need to be kept updated on a regular basis so
the duties are current. As product-line changes or machine changes take
place, it's reasonable to expect some change or adjustments will need to
be taken to fine-tune the job description. Employer, manager, and
worker need to have the same perception of what each job in a factory
requires.

Managers and Authority

Some managers are not willing to exercise authority because they
may be timid or unsure of themselves. Not everyone is able to exert the
appropriate amount of authority to address some specific issue in a way
that brings credit to himself, the company, or the subordinate employee.
Whatever the reason or reasons may be, not everyone is going to be a
good manager.

Just for self-evaluation, managers ought to take the time to look at
textbook descriptions or elements of supervision. This is not to say the
reader has to agree or believe all authors have answers suitable for his
or her own situation. It's useful, however, to remember what some specialists claim management's duty and obligations are for the individual
to assume, along with title and reward.

Organizations, regardless of size, should be able to identify position
and responsibilities of each participant taking part in the operation of a
company as stated in the previous paragraph. Each position should have
a job description, which lists the responsibilities for an incumbent to
carry out while in the specific assignment, assigned to the work slot. The
CEO, who would be responsible to a board of directors, is not exempt,
nor are board members. The position description is kept up-to-date and
is readily available and given to the individual in the position.

A table of organizations for industry or the military is the index of
the operations from which staff and outsiders look to seek information.

A flow of authority can best be described by asking some questions such as:

(a)   What are we going to do?
(b)   How are we going to do it?
(c)   Where are we going to do it?
(d)   When will we do it?
(e)   Who will do it?

These questions form a basic network of duties that face each supervisor.

While boards of directors examine organization goals that deal with growth acquisitions and financial stability, the chief executive officer establishes policy to regulate the business and is empowered with authority to achieve corporate targets.

Supervisors, whose fundamental duties are to organize and to work through others, are expected to carry out executive plans of work to reach the goals.

Managers need to remember that they must be able to show the leadership that gets results by using authority with a series of common sense methods. If managers fail to create a process of employee responsibility, then it's more than likely the organization is destined to encounter problems.

This idea of employee responsibility is a two-way street. Managers cannot expect the employee to assume responsibility of waste if the worker is given poor equipment or bad raw material from which to produce products.

One of the critical work failures I see taking place in food plants goes to the very heart of job descriptions. I sometimes talk with line workers and ask what their job entails. Later I ask the employee's supervisor what the worker's job is supposed to be. You would be surprised to find out that there can be meaningful differences in what the line worker believes the position requires, versus what his or her supervisor thinks the work station requires.

Videos and Other Electronic Gadgets

Purchasing personnel need to be instructed that when new equipment is being purchased the supplier is expected to provide instructional video training tapes showing how to operate and maintain equipment being acquired.

Existing operations in a factory need to be filmed with players who are qualified in showing the functions and expectations of the unit operations. The video ought to show equipment dos and don'ts of a unit operation. Information from startup, run, shutdown, cleanup—the full cycle—has to be orchestrated.

What this means is that management can develop or acquire a library of videos it needs to train staff. The library safeguards operational procedures in such a way that the uniform operational protocols can be transferred from worker to worker.

Not without importance is that, when videos are made, the line worker and the supervisor soon see habits that have been acquired that often reflect, "I didn't know I did that," or flaws show up that may well have not been recognized.

Long-Term Training

Training to teach tangible skills and the tricks to develop long-term beneficial attitudes between supervisors and line workers is an investment in human capital that becomes the very backbone of an organization. For years I would lecture that for managers the idea is that "if you care, they care." It takes both the verbal and physical commitment of senior executives in organizations to set the tone or philosophy that people respect to create successful organizations.

When the worker approaches the supervisor and tells the person that something is amiss in the process, be it a machine or product material problem, he or she needs to be listened to. If you show that you apparently do not care, it's more than likely the employee will not tell you again. There's a lot of room and reason for supervisor to pay more attention to people on-line. The "hands-on" supervisor still has critical value in food-processing installations. There's still a lot of value in going up to people operating food-processing equipment and asking for suggestions on how to improve product flow-through rate or how to make sure defective material is not put into finished-goods market items. Of course, many suggestions may not be feasible at the time, but it's good practice to respond with rational reasons why even though the idea is useful, the firm is unable to correct the problem at the time. The supervisor needs to keep productive dialogue flowing between staff.

Production Management Training

In some parts of the globe we see that managers or midmanagers are required to demonstrate well-developed technical skills before the individuals are allowed to progress within corporations. What is not readily visible is that while a manager may be technically competent, no case can be made that the manager needs to demonstrate "people management skills." I've been told on many occasions by on-line staff in food factories that immediate supervisors are reluctant to speak out against work plans that generate waste because they might be blamed for the poor operating practice. It almost seems an individual department needs to protect its own "turf" without calling attention to individual problem zones. While chief executives of small and big companies are supposed to be able to provide the vision needed to keep their organizations healthy, we see the role they play is not transferred to first- and second-level supervisors. Little to almost no effort is being made to train undereducated or undertrained individuals holding down supervisory positions to do a better job.

Managers' human resources, unlike robots, possess drive, energy, and smarts, along with the many attributes they are supposed to possess. While managers are supposed to carry out company directives and to support policy, they need to have vision about the future of their organizations.

It is not enough to tell managers "you just take care of your end of the job" and leave the more "complicated" activity to top managers. Remember, first-line supervisors, along with shift managers, convert organizational strategic plans into reality, and they, with the on-line workers, make the action plan take place. If we fail to provide continual training to keep skills sharp then we can be criticized for not portraying commitment to excellence. Just as equipment needs maintenance and periodic overhaul, so do people who operate equipment. Breakdowns occur when machinery is not lubricated; so will the manager who is not plugged into the "vision, creativity" loop.

Critical Time

Large companies find it easier than small firms do to find time slots to train staff, because more people are available to pinch-hit for persons pulled out of a work assignment. Training costs are not cheap, and can limit additional education for organizations that feel the training excuse is not a budget item in their plan of work. Training, be it formal or not so, can be ongoing if one or more individuals are prepared to seek information and then transfer the information one-on-one to personnel in the factory. Sometimes the people who do attend off-site training programs at colleges or in industry meetings fail to transmit the information back to persons in their organization who can benefit from information or literature picked up at the meeting.

CHAPTER 11

Unconventional Techniques to Deal with Waste Recovery or Treatment Schemes

Oily Wastes

Not too long ago, a food plant executive called me for advice on how to save money in treating oily wastes being discharged from his factory. The consultant engineer recommended pretreatment grease trap hardware to meet fat and oil limits set by the local sewer authorities, who were adding surcharge costs to the firm's treatment billings. The surcharge costs amounted to many thousands of dollars per year. The engineer's estimate to build a pretreatment system was about $100,000.

What would you do?

Civil engineers tend to think in concrete and structures, and may not think in terms of product chemistry or select physical characteristics of the problem material.

My approach was to ask the executive whether or not the firm had cooling equipment. The answer was affirmative, and my next query was whether there was sufficient refrigeration reserve for cooling wastewater, which I presumed was warm to hot. The response to my question was that yes, refrigeration was available, and how did I know the waste flow was hot? You might say it was a lucky guess, but in truth it's more like I made an educated guess than just a shot-in-the-dark question.

My advice was, "If you can get a representative composite sample of the waste discharge, let's cool it to some point where the fat components congeal and can be skimmed from the liquor fraction. Try that approach and call me back." The next day an excited owner called and reported that by just cooling the waste stream 10°, the fat fraction was solidified and could be removed.

What would you suggest as a next step?

My approach was, "Have management intercept hot waste from unit operations losing fats at the source before being discharged to a sewer drain and chill the fats to recover edible fraction for reuse. Call me back if it is a workable idea."

In a couple of days the call came back with a more-or-less typical response. *"Do you know how much money you saved this operation?"* Of course this was a success story, but not all attempted solutions are such. We all have failures too.

Clam Wash Water

Another unconventional solution to a problem involved recovery and use of clam wash water. A team of professors (including myself) and graduate students from Cornell's Department of Food Science were studying the practices of the seafood industry when an opportunity was noticed.

I noticed that, when the surf clam was washed, the water became foamy. The suspect cause of the event in my mind was that foam probably originated from undenatured protein material lost in the water. When the clam was minced (cut into strips), small particles of meat were dispersed to the water carrier. In my mind it seemed reasonable to think the wash water could be used to produce economic credit to the factory. Figure 11.1 shows a flow diagram of a surf-clam-processing operation.

One of the products the clam-processing factory produced was canned clam broth that was sold to institutional trade for making clam chowders. The basic protocol to make the clam juice or clam soup was to cook clam meat in water, add salt, can, retort, and label. The meat used to make the broth had little value after cooling and was separated as waste.

Clam meat, the main product in the factory, was obtained by shucking clams obtained from ocean beds, washed, minced (cut into strips), and sold to restaurants like Howard Johnson.

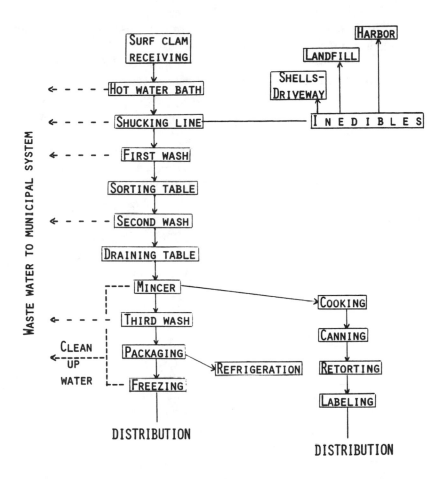

FLOW DIAGRAM OF SURF CLAM PROCESSING

Figure 11.1. Flow diagram of processing steps used in a clam operation.

The minced-clam operation, together with clam washing operations, generated three different streams of wash water, as shown in Figure 11.1.

The last wash station generated wastewater with a BOD ten times higher than the first station (234 to 2340 BOD5).

The final story turned out to be that the third-wash water was intercepted in sanitary equipment, boiled, and concentrated about 10 percent, salt-added, canned, retorted and labeled. Of special interest was that the broth, made from the mincing operation, was judged better than broth made from cooking clams.

You may well imagine the stir I caused at a National Food Waste conference when I reported on the clam project as I flashed a flavor evaluation of clam wash water on the screen. Yes, we trained ourselves how to judge wash-water flavor to duplicate a product. The evaluation form is shown in Figure 11.2.

The engineers were never going to taste sewage (sewage it would be, if the wash water was sent to the drain). There's a full report about the project available in the 7th National Symposium on Food Processing Wastes, Atlanta, Ga. EPA tech series, EPA 600/2-72-304 pp. 42–66 December 1976.

This part of the overall project changed, to a large extent, the way clam broth would be produced and canned in the industry. As to the effect of removing the wash water from the sewer, the BOD load was reduced by 40 percent.

Fish Scales

The author of a 1991 report concerning how to improve the profitability of fin fish-processing waste cited the work of one of my graduate students and myself, in which we showed that fish scales are useful as an edible coagulating ingredient to purify wastewater.

The thought to develop this idea came from chitosan publications where different researchers showed that a 75 percent recovery of chitosan was extracted from shrimp, crab, krill, and other seafood waste.

When asked by a colleague whether or not I thought we could do something with fish scales, it seemed to me that scales might be used like chitin. Chitin can be extracted from a variety of sea creatures; the conversion of chitin with sodium hydroxide can make chitosan. Chitin then can be used as a coagulating agent in treating waste.

Fish scales are not minor components in fish-processing wastes. About 2 percent of fish weight is attributed to the scales. One author writes that a firm processing 50 tons of fish per day needs to deal daily with a ton of scales.

Increasingly stringent pollution control measures have forced seafood processors to by-product recovery schemes; therefore, getting rid of scales is important. Fish scale can be a useful feed by-product, especially when 10 percent of it in a diet can be consumed by poultry in

Tasting Session I

FLAVOR EVALUATION OF CLAM WASH WATER

Directions: You have before you 10 samples of processed clam wash water and a sample of canned clam juice "R".

Taste the reference sample "R" first and evaluate samples 1–10 against it. Samples 1–6 are unsalted so taste these first and then proceed to 7–10, the salted samples.

Flavor attribute	Intensity rating									
	Much less		Mod. less		Slightly less	Refer- ence	Slightly more	Mod. more		Much more
	-5	-4	-3	-2	-1	0	1	2	4	5
Clam flavor										
1										
2										
3										
4										
5										
6										
7										
8										
9										
10										
Fish flavor										
1										
2										
3										
4										
5										
6										
7										
8										
9										
10										

Figure 11.2. Work sheet used by a research team to determine a flavor profile of clam wash water.

meal-feeding programs. Of course, the dollar return from taking this path is not lucrative.

It is possible to increase dollar value return by using fish scales as a coagulation chemical produced with an operation that needs very little capital. Fish scales can be air-dried and then simply milled to pass through a 0.5 mm screen.

Fish scales are made up of about 33 percent mineral with significant content of proteins. These mineral/protein-containing fish scales can be prepared for use as coagulants by dispersing the powdered scales in water as 0.01 percent solutions. Our experience with this material was obtained by using it in waste streams obtained from egg breaking plants, scallop-shucking operations, and fruit-juice processing. Coagulant additions at 15–25 mg/l, along with pH adjustment to waste streams, reduced suspended solids 95–96 percent. F.W. Welsh and R.R. Zall reported the work in *Process Biochemistry*, August 1979, pp. 23–27. The value of using scales as a coagulating agent can be worth more than just a supplement in animal feed. This information, along with reports of others showing the value of chitosan, leads to other uses for fish scales.

Another graduate student and I looked at using a membrane cast with chitosan and fish scales to treat wastewater containing heavy metals. Absorption techniques using natural polymers such as chitin and chitosan were suggested by others in the early 1970s. T. Yang and R. R. Zall published results of their work in the *I & EC Product Research and Development* 1984, 23, (1) 1988, published by the American Chemical Society. Fish scales were found to be superior in the chelating of metals to the chitin and chitosan material.

It seems to me that we may well be overlooking more valuable uses for seafood wastes than what has been going on in this time frame. My students and I have been awarded patents for producing useful by-products from seafood wastes and other commodities. Substrates like clam bellies and fish innards contain a myriad of enzymes useful for tenderizing meat, making detergents, and even replacing rennin in cheese-making. In my opinion, we have been reluctant to look at unconventional schemes to deal with waste problems/opportunities.

Egg Breaking Wastes

No other commodity that I've worked with provides more value to the idea for segregation, separation, and fractionation of waste flow than does this material.

Egg breaking losses are enormously high because

1. The avian scientist went on his or her way to increase egg production per bird from about 120 eggs/bird/year to over 200. A negative value of this

genetic-breeding accomplishment was that eggshells became fragile, were less able to resist damage, and were easily broken.

2. Hen housing husbandry switched from letting birds run loose in pens to keeping them in wire cages. Results of the change meant that eggs produced by hens in cages were relatively free of manure and really didn't need brush cleaning with water before being broken as mandated by USDA rules.

When fragile eggs are mechanically cleaned with a brush, a large number of eggs will break in egg-washing machines, causing losses of 6–12 percent. Given this information, we recognize we have a high BOD waste stream rich in organic loss. This is the bad part; but, on the plus side, egg-washing operations produce only 15 percent of a total loss volume sent to sewer drains in egg breaking operations.

Does it not make sense to separate the egg-washing liquids from relatively "clean" wastewater? We have stressed that any plant manager worth his or her salt would intuitively separate clean wastewater from dirty wastewater in a food-processing factory to make waste treatment easier. This step provides other choices.

One less expensive way to treat high BOD wastage in operations like the one being described is to collect the effluents from egg-washing machines, then cart the volume in tank trucks to public municipal treatment plants and pay for wastage like septic-tank waste products. There are people who provide this service.

However, there is still another choice if the solution fits the factory. Just as eggs are part of nutritious diets for humans and other mammals, egg solids can be used for fish food.

In 1981 I worked with people in the U.S. Fish and Wildlife Service, and we tested fish food made from egg wastes and dairy-processing wastes. Wastage from both industries was concentrated and deep-oil-fried to yield products that could be encapsulated for fish-food feeding trials. The waste provided both fat and protein for rainbow trout diets. While the dairy waste replacement in 10–20 percent amounts in fish food resulted in higher body weight gain per gram of food fed than control feeds, egg wastes were far better.

The trials with rainbow-trout-feeding with egg breaking wastes produced almost a 1-gram weight increase per 1-gram feed, which, as I understand it, is remarkable in animal/fish nutrition.

Well, then what is the claim? Managers are able to segregate and concentrate food processing wastage in low-volume water, carrying amounts we may be able to reuse, recycle, or reduce treatment problems.

Reclamation and Hot/Cold Water Use

To those organizations which use evaporator concentration systems to condense liquids or food slurries, there is the opportunity to salvage hot vapor with almost distilled-water quality. A world-proven example is the recovery of milk vapors in powdered-milk operations. To my knowledge, I invented the process around 1953 in Arkport, New York, when I managed a large multi-product milk-processing plant. We processed a million pounds of milk or more daily; most of the milk was condensed prior to powdering. Skimmed milk was concentrated 5 to 1 in evaporators before drying, which generated about 700,000 pounds, or about 80,000 gallons, of clean, hot water. I coined the term (perhaps not the best one, as I think back) cow water, which now identifies the by-product around the globe. The method used to capture the hot water is shown in Figure 11.3.

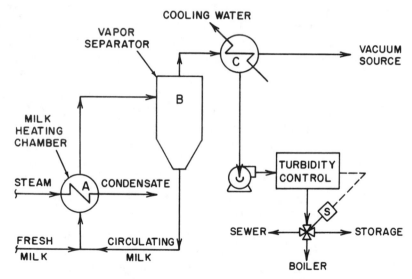

Figure 11.3. Recovery of milk vapors in powdered milk production. The milk evaporator functions much like steam generated in a boiler. Because the vacuum pan is subjected to a high vacuum through the air ejector system, milk boils at low temperatures that rarely climb higher than 150°F. Milk vapors change back to water, which collects against the cold condenser tubes. Vacuum increases linearly to maximum with lowest temperature; thus, vapors flow from A to B to C. Malfunctioning milk evaporators tend to foam, carrying over milk solids; hence, an in-line turbidity meter safeguards the condensate purity for the salvaged water reuse.

Prior to the salvage scheme this water source was commingled with sewage effluent that complicated waste treatment. Later on after salvage, unused segregated hot water was mixed with other clean water and atmospherically cooled in open channels prior to being discharged directly to a trout stream.

Some of the hot water was used for cleaning, for boiler feed purposes, and for a source of heat in heat exchangers used in making ice-cream mix. Some of the water was cooled and packaged in attractive blue milk cartons and sold for use as soft water for batteries and hair-washing. No claim was made at that time that the hot vapor was a drinking-water-quality product. Needless to say, the salvage idea was picked up by machinery manufacturing organizations, and C .E. Rogers manufactured and sold recovery units to plants around the world.

What is of special interest to me with this water salvage operation is that factories burdened with a poor water supply, or even with a lack of adequate water supply, can use the water salvage system as a primary water source. Plants in Vermont and Wisconsin with evaporation operations, for example, were built and run with a water supply extracted from concentrated milk processes. The Vermont site lacked water; the Wisconsin location has a poor water supply with a hardness content of almost 340 ppm or 20 grains.

Sometimes it is useful too to look at a variety of water conservation schemes by accumulating seemingly small volumes of water into a single tank and then adding fresh water as needed to provide amounts needed in a factory. At one location, the following scheme was used and proved to be both useful and effective. See Figure 11.4.

Very often, a single-pass condenser, like those often used for air conditioning or other small heat-exchange units, cooling water can add up to hundreds of gallons of cooling water/hour in some factories. This approach will assist water conservation and lessen wastewater volumes to be treated.

Membrane Systems

Just a few words about membrane systems—some of the early work with food systems was looked upon as "mad scientist at work: beware." Concentration through ultrafiltration and reverse osmosis made its debut by treating brackish water with excellent success. Further uses of the process are in place in assorted industries. Food by-products, such

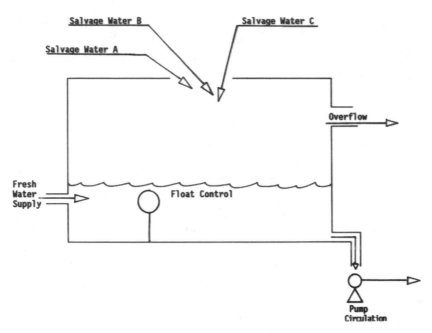

Figure 11.4. Water recovery tank which accumulates recycled water and supplemental fresh water.

as cottage cheese and whey, are being processed through appropriate membrane material. Proteins, which are large molecules, are captured in this system by using molecular sieves with pore size in the range of 20 angstroms. Macromolecules in solution can be screened almost independently of pressure with high flux rates (the solution permeating a membrane) because ultrafiltration is a function of Reynolds number and solute concentration with little pressure drop.

Reverse osmosis differs from ultrafiltration by its ability to screen smaller molecules, such as dissolved salts. In a reverse-osmosis system, flux rates increase linearly with pressure. The principal ingredient in whey is lactose, a milk sugar that is a much smaller molecule than are proteins, and it can be concentrated prior to salvaging by reverse osmosis. One of the spin-offs of using membranes to fractionate and concentrate liquid foods was a low BOD water stream. In 1969–71, when I was project director for the EPA's funded first food-membrane-processing operation, we recognized the value of using RO (reverse osmosis) water for cleaning purposes. Suitable water quality is mandatory when cleaning membrane systems, especially the cellulose acetates and many of the polysulfone varieties. Figure 11.5 shows a two-step cheese whey fractionation system of the Crowley Milk Project where the first use of RO water was returned for washing cottage cheese. Later on in other operations RO water is used in cleaning systems.

With great reliability achieved from membranes during the 1971–80 period, it seemed possible that RO water might even be used as beverage material. Select data from a bulletin of the International Dairy Federation N232 are excerpted from a booklet. Tables 11.1, 11.2, 11.3 and 11.4 are reproduced along with some legend information.

From the article the following is extracted: "RO membranes have not only a large retention level of organic materials, but also a large retention of mineral constituents."

The author measured the level of selected minerals in retentate and permeate from an RO plant processing whey. I compared these levels with those found in various commercial mineral waters. The results are shown in table 11.4. RO membranes have not only a large retention of organic material, but also considerable retention of mineral constituents.

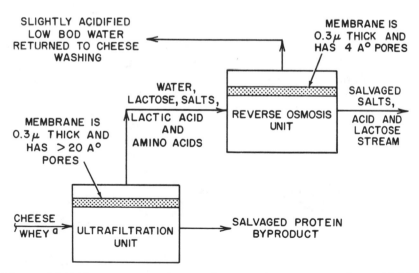

Figure 11.5. Two-step membrane plant used to process whey with UF and RO systems.

Table 11.1. Select test data derived from processing sweet whey

Description	Units	Specification
Feed		
Type	—	Sweet whey
Total Solids	%	6.0
pH	—	6.0
Temperature	°C	30.0
Hydraulic rate	l/h/module	448
System flux	$l/m^2/h$	16
Concentrate		
Temperature	°C	30.0
Hydraulic rate	l/h/module	111.0
Total Solids	%	24
Permeate		
Hydraulic rate	l/h/module	337
Hydraulic flux	$l/m^2/h$	12

A Toxin-Free Water Supply

What makes the water extracted from milk/milk products especially interesting is that the cow is a biological filter that removes toxic elements from food and water it consumes. Metals accumulate in the animal's bone system and will not pass mammary glands to contaminate milk. With more emphasis being directed to "pharmacology"—that is, using cows to produce drugs—it seems that using cows as toxin filters makes sense too.

Table 11.2. Membrane retention data

Parameter	Raw Whey (mg/l)	Permeate (mg/l)	Retention (%)
COD	64,000	850	98
BOD	29,500	430	98.5
Total solids	62,400	1500	97.6

Table 11.3. Comparison of mineral retention of water from different sources

Products	pH	Minerals (mg/l)				
		Na	Cu	Ca	Mg	K
Perrier water	5.0	12	0.009	137	3.4	0.6
Vichy water	6.3	1090	0.02	93	9.5	70
Vittel water	7.7	3.8	0.01	180	3.7	2.6
Deminerized whey	8.0	13.6	.003	0.14	0.01	0.65
Non-demineralized whey	8.0	69	0.006	2.70	0.71	193
RO permeate		60	0.006	20.0	0.50	160

Note: RO is reverse osmosis.

Food from Waste

Not quite 30 years ago, a book titled "Food From Waste" was edited by G. G. Birch, K. J. Parker and J. J. Morgan with views expressed by 25 scientists working in this area. The text contained 20 papers presented at an industry/university symposium held at the National College of Food Technology, University of Reading, England.

Table 11.4. Select constituents present in permeate from an RO system processing sweet whey

Constituents (mg/l)	Permeate		
	Day 1	Day 2	Average
Ash	4900	4900	4900
Minerals			
Calcium	296	305	301
Magnesium	75	75	75
Phosphorous	442	391	417
Potassium	1550	1460	1515
Sodium	378	414	396
Zinc	0.5	0.5	0.5
Iron	2	2	2
Vitamins			
Thiamine	0.15	0.37	0.26
Riboflavin	3.5	3.6	3.6
Amino Acids	530	510	520

Subject matter included conversion of fungi to food, production of microbial proteins from carbohydrate wastes, food from waste paper, proteins from starch mill effluents, etc.

Considerable overview information and ideas were generated from this symposium and others like it over the following few years.

What struck me as a real issue was a statement put in print on the slick outside paper cover which read as follows: "In the next two or three decades production of food from waste will become imperative, and no source of nutrient, no matter how disgusting, can be excluded from consideration."

Of course the kind of waste to be converted into foodstuff will depend on location needs. For example, Israeli dairy farmers are known to separate undigested feed grain from cow manure to reuse for non-milking animals.

A dairy farmer in California uses methane from cow manure to power a generator, which produces electricity that is then sold to the Pacific Electric & Power Corporation. Spent manure liquor is fractionated into protein and cellulose material. The cellulose fraction is dried and used as cattle bedding. The protein liquor is fed to calves.

What was not an agenda topic 20 years ago was the use of microorganisms with altered genes that could be used to produce products of value. The press and television frequently report about breakthroughs in this area. We now have FDA-approved rennin substitutes as one product, and there are others. As yet, genetic engineering is only being directed into high-value areas. It's reasonable to believe that, in time, the gene manipulation process will be used to mutate a variety of microbial species that can be used to take apart and convert agricultural waste which, to date, is underutilized.

Salt Whey in Egypt

Sometimes managers need to step back from a problem to better understand either its magnitude or its solution.

I was invited to Egypt a few years ago to take a look at educational programs and industry practice. One of my duties took me to an Egyptian Army camp where a Brigadier General commanded food production operations. The army, for example, bakes bread and other commodities that are sold to citizens. Because of my dairy industry experience, I was shown a turnkey cheese operation built by the Danes for Egypt that made feta and other soft cheese products. The general was looking for a system to treat salt whey being discarded from the cheese factory.

You may not know that 6–10 percent salt by weight is added to non-beverage milk in the Middle East to prevent souring. When the salt milk is made into cheese, the whey fraction is troublesome.

In fact, there was talk of giving a prize to the firm or person who could come up with a solution for what he perceived to be a difficult problem. Let us forget for a moment that whey proteins could be separated or that lactose might have value. The problem was complicated by availability of funds.

The army base is located in a rural desert area. Salt whey quantity amounted to just over 30,000 gallons/day. As I saw the issue, the solution was well within the reach of the local commander and with little cost.

My suggestion was to alternately gravitate feed salt whey to four or six separate one-acre lagoons. The lagoons' depth ought to be 2–4 inches. The rationale is that an acre-inch will accommodate about 27,000 gallons of liquid, so each lagoon would take care of a day's salt whey.

Heat and sunlight, a blessing and a curse in Egypt, would evaporate the salt liquid in 1–2 days. (Seepage into sand would no doubt take place until sand pores were filled with whey debris.) Dried whey/salt solids could be shoveled up and might be useful as a feed supplement.

Somehow the problem was exaggerated because some on-site engineers wanted to treat salt whey (which is a difficult material) using conventional waste treatment hardware. The staff failed to see its resources of land/dry air/heat as valuable assets. I do not make light of waste problems, but I think some engineers tend to want to build monumental treatment plants. Plant design, some engineers feel, shows the expertise of the individual.

Raising Eyebrows and Monuments

I raised a few eyebrows with some of my views at an international conference in Ireland on dairy effluents. My commentary included a statement that "the sewage handling activities in milk plants in Ireland appear to be monuments to waste." With due respect to sanitary or civil engineers, it's probably correct to suggest that some treatment plants can be overdesigned just as well as being underdesigned.

When one goes to a country or region, one has to understand why overdesigned conditions might occur. In Ireland it's to a firm's advantage to treat waste to almost drinking-water quality. Part of the system's construction costs and annual operating expenses of the sewage treatment plant are returned to the company by the government when the system's discharge is within water quality guidelines. The Irish want to protect their waterways for fishing, as well as the environment for hiking. Ireland's nearly pristine beauty is the attraction for tourism, and an unpolluted environment is a major source of revenue.

Winter Coldness for Refrigeration

Waste-management schemes provide "seeds" to look at other conservation practices in food factories. Just as the idea of using sludge to generate methane to burn as fuel or produce electricity, we may have other assets. We can conserve, create, recycle, or become more efficient by working with natural resources.

Considerable work has been undertaken to convert wind power to electricity with some but limited success. More benefit has been obtained by using solar energy to heat water, but not much to generate electricity.

What about cold weather? About 25 years ago I published my first report about using winter coldness to provide refrigeration. The idea was a takeoff of solar heat, but rather than to heat water, cold air would be used to chill cold rooms, cool water to produce ice, and chill antifreeze-protected water to circulate as a coolant liquid. Over a 3-year period it was established that in upstate New York it was energy-efficient to use cold-weather air almost 5 months out of the year for chilling purposes. In fact, ice could be produced for as little as 2 cents/ton during much of the winter time. When outside air dropped to 26°F or below, factories could chill cold rooms with unlimited amounts of cold air supplied by low-power fans and exhaust dampers.

Is My Air Cold Enough?

One might ask, "Is the weather in my part of the country cool enough to use as a refrigeration resource?" To quantify the cooling power you have available, simply obtain degree day information in a given month and year from your local weather bureau or utility company.

The term "degree day" is used to calculate energy savings and is defined as the difference between 65°F and the average of the high and low temperature in a given day.

The sum of degree-day numbers becomes an index. The higher the number, the better, because a firm would theoretically need more fuel to heat its plant. The heating degree day (HDD) is a climatologic statistic that is useful to a variety of users like contractors, architects, etc. Negative differences count as zero.

Table 11.5 shows some heating degree-day numbers in different cities of the United States.

Table 11.5. Heating degree days in representative cities in the U.S.

Birmingham, AL	2780
Boston, MA	5791
Tucson, AZ	1776
Detroit, MI	6469
Los Angeles, CA	1451
Minneapolis, MN	7853
San Francisco, CA	3421
Winston-Salem, NC.	3595
Denver, CO	6132
New York, NY	4965
Washington, DC	4333
Cleveland, OH..	6006
Jacksonville, FL	1113
Pittsburgh, PA..	5905
Atlanta, GA	2811
Memphis, TN.	3137
Plains, GA	1895
Houston, TX	1388
Chicago, IL	6310
Dallas, TX	2272
Des Moines, IA	6274
Richmond, VA	3955
New Orleans, LA	1317
Seattle, WA.	5275
Portland, ME	7681
Milwaukee, WI	7205

Figure 11.6. Four fin radiators (actually truck-type) arranged in a square with a low-watt fan mounted in the center to draw air across heat exchange area.

There seem to be considerable opportunities to convert cold weather into a usable resource. Bear in mind that most of the food industry operates in cold climates. It's true that refrigerated warehouses operate in cold countries too. It seems as if we truly have only touched the tip of the iceberg.

I've been blessed with industry support to fund some of my antics over the years. New York State Gas and Electric took a lot of interest in my schemes to recycle cleaning liquids as well as the idea to utilize winter weather.

You may not know that utility companies have a keen interest in future power use. They welcome ideas to use less energy as waste so as to

Figure 11.7. Cabinet cooler (heat exchanger) where water cascades down over the exterior; air-chilled propylene-glycol pumps through the interior.

save the costs of building power plants. Probably also not well known is that they, by law, spend a percentage of their revenue by funding research.

One of the demonstration projects New York State Gas and Electric and I put on-line in late 1989 was a chilled water plant for the operator of a poultry-processing plant.

Propylene-glycol (food-grade) was used with a water solution and circulated through fin radiators mounted on the factory's roof, where the water was chilled by outside air.

Chilled coolant was then circulated through a surplus cabinet surface cooler that was used to chill water pumped over its exterior leaves at 1000 gallons of water per hour. Water was chilled from 55°F to 32–33°F and pumped to three food-grade plastic storage tanks. Storage capacity amounted to 10,000 gallons.

Figures 11.6, 11.7, 11.8, and 11.9 show the hardware. Students and their instructor are shown visiting the site as part of my waste-management course. The economics of cooling water with outside winter air can be very rewarding, especially if little pumping energy is used.

Figure 11.8. Overview of the air-cooling fin radiators mounted on plant roof with three food-grade plastic tanks set in place outside to store chilled water.

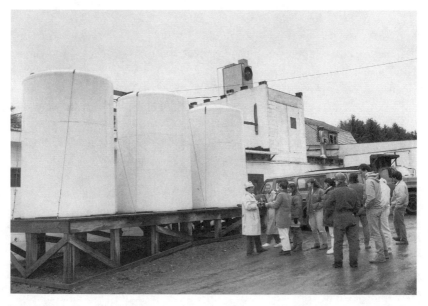

Figure 11.9. Students and their instructor on a field trip to review waste and energy conservation methods.

The trick to the conservation methods is to look at where it is best to locate heat-exchange cooling units. If I were to place the system described in factories of choice, I would not wish to pump liquids any more than needed. Where possible, we need to remember the value and low cost of gravity flow.

CHAPTER 12

Layman's Overview of Treating Water, Wastewater, and Solid Waste

In some parts of the globe, drinking water comes from the tap; wastewater goes down a drain; human wastes disappear with a flush of the toilet, and solid wastes need to go into a dumpster. Most of us in Western society have never been made aware of what it takes to produce potable drinking water and what effort goes into treating our liquid or solid waste.

The key concept that this author thinks needs to be stressed is that humans and their progeny contaminate and pollute the air, water, and soil of our Earth by existing. The degree that contaminates occur will vary, but to remove the pollutants means we need to separate, fractionate and then concentrate pollutants into some form that can be recovered and eliminated. Treating systems we are forced to build render pollution harmless, when properly constructed. Figure 12.1 shows a combined sewer system that treats domestic, industrial, and storm-water waste.

Water Supply

Not all food processors view water quality the same way. The fact that water appears crystal clear does not mean it's safe, even though clear water gives consumers an indication of its acceptability. The ideal water supply is one that is "soft," cold, and free from impurities, but is

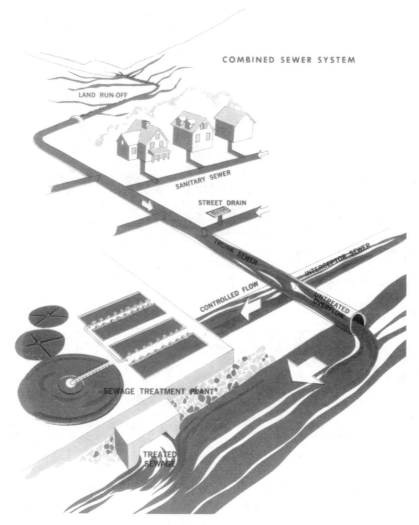

Figure 12.1. The combined sewer system includes connecting and trunk lines that carry waste to a treatment plant. In times of large rainfall, raw sewage overflows into a receiving stream or larger body of water.

rarely available in enough volume to satisfy a food factory. Parameters other than turbidity, taste and odor are important depending on end use. Water needs to be safe microbiologically as well as chemically.

Water, when used for cleaning purposes, probably needs to be softened, unlike water used in flumes that convey fruits and vegetables in canning plants.

Water conditioning, when required, is not inexpensive, and where conditioned water can be recycled there are potential in-plant savings. Recycling provides additional savings by decreasing water used by the factory that is sent to waste treatment. Water is a two-sided sword that costs money to acquire and costs money to discard.

From the author's experience, at least 50 percent of water used in a food factory can be reduced when managers put their energy into devising systems to curtail water waste.

Let's not leave the issue of water quality without noting that soft water has a down side because zero-hardness water is corrosive. Water, when softened with zeolite systems, can cause hypertension to people drinking such water. True, too, is that kidney stones can be induced in people who consume hard water. What ought to be stressed is that it behooves factory people to learn about what kind of water quality a factory needs and what must be done to treat it wisely for use within the site utilizing the commodity. Managers, who have operations that use water, need to know how best to use, conserve, treat, and dispose of water in preferred systems appropriate to the processing system.

Wastewater

First of all, most people haven't a clue that liquid wastewater from a community is essentially the water supply being used by the community. Wastewater will probably be 99.9 percent water with as little as 0.1 percent fouling-material content. In general, domestic waste strength approximates 250 mg/l BOD5. Across the board, food plant wastes probably contain 2500 mg/l/BOD5, or 10 times the concentration of domestic waste. This means that plant wastes contain about 99 percent water with organic wastage of about 1 percent.

Way back at the beginning of this book I described the concept of separation, fractionation and concentration as what needs to be done to treat waste. The solution to pollution is not dilution, but concentration. When we waste fewer fluids the probability is that we can lessen treatment costs by:

(a) Waste treatment systems can be smaller (less cement, lagoons, aerators, etc.)—

(b) Energy use is less—

(c) Land mass to construct plants could be less—

(d)   Operational costs may be less—
(e)   Plant wastage could be less—
(f)   Product yields may increase—
(g)   Profits may be greater—

Of course we could create larger benefit and problem lists; but, in general, the more we decrease waste flow, the better we can deal with waste-handling situations.

Food factory wastes can be relatively soluble or insoluble in water used to flush the wastes to sewer drains. Insoluble wastes are easy to see and can easily be screened from the liquid carrier by filtering or settling. In fact, dry cleanup activities in a plant are cost-effective actions because we need not build a treatment system to separate solids from water carriage at the treatment plant.

Soluble wastes that cannot be physically separated need more expensive treatment. In such situations, our engineers use microorganisms to convert organics into biomass. When the biomass or floc is of sufficient size, they are separated by physical means into material people call sludge. Pondlike structures of varied configurations can be constructed to create very big broth tanks where bacteria multiply by eating the waste (food) and are then separated as the pollutants for further treatment.

Of course the treatment systems are used to separate, fractionate, and concentrate the waste into some final condensed by-product that then needs disposition. Landfills, sludge digestion systems, and others are used to still fractionate wastage into some manageable condition.

There are lists of "dos" and "do nots" put out in print by university extension agents, the Environmental Protection Agency (EPA), and private engineers who list dry cleanup activities as well as related activities. What all these recommendations come down to is a list of "management" responsibilities. I'm reminded again of the axiom taught in my classes over many years, as well as in industry where I managed food plants, which is: "if management doesn't care, the employee doesn't care." If you walk over a running hose left on the floor by a careless worker without corrective action, the employee is apt to think you do not care about the loss and complication of wasted water.

Perhaps a more difficult idea in waste treatment protocol is an understanding of the biological systems needed to take care of dissolved organics. In general, the systems for treating wastewater and its sludge by-products are aerobic or anaerobic. The aerobics generate little odor when operating properly. The anaerobic systems are odorous, as they can produce sulfur odors like rotten eggs and a myriad of ill-smelling rotting odors associated with decayed meats, etc.

The basic biological processes are not that many and can be defined by:

1. Presence or absence of dissolved oxygen (i.e. aerobic or anaerobic)
2. Photosynthetic ability
3. Mobility of organisms (i.e., suspended or adhered growth)

General Waste-Treatment Primer

Man fouls his environment by concentrating large numbers of people in a small area or by creating industries that discharge waste into a confined location where nature is unable to purify the waste.

By purifying waste pollutants, both organic and inorganic material are broken down into simple compounds that lose identity and will no longer deplete oxygen or dirty water.

Waste purification that takes place naturally may take days, weeks, or months. The purification stages may be aerobic or anaerobic and involve single-celled or multi-celled organisms. Bacteria assist purification as they consume waste as food so they can reproduce. Microbial cells themselves become food for other "critters" and so on until oxygen reappears in the watercourse along with fish and protozoa. By the process, water is once again clean.

Figure 12.2 illustrates the more-or-less natural purification stages in a water flow.

In order for a large number of people to live in a confined environment, man has had to devise waste-treatment systems that speed up natural purification processes to take care of waste generated by people and industry. Wastewater can be defined as domestic, street-washing, storm-flow, groundwater and industrial waste.

Industrial waste can be further broken down into waste from chemical, steel, food and kindred products, etc. Regardless of source, basic methods are used to deal with wastage. Processes used to treat waste consist of physical, chemical, biological, and combination techniques. Waste that enters a treatment scheme is called influent, and the discharge is effluent. The solids in wastewater can be divided into two general groups: organic and inorganic. Organic solids originate from animals and plants and will decay over time. Inorganic solids will not decay. These solids are suspended or dissolved in a water carrier.

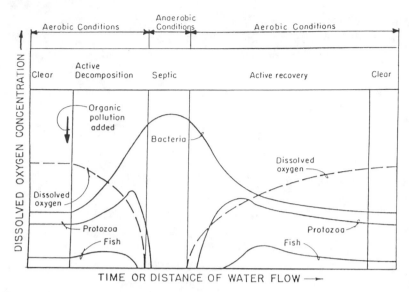

Figure 12.2. Wastewater degrades naturally over time when discharged to a watercourse with sufficient assimilative capacity to absorb the organic load.

Treatment

Because sewage solids can be both suspended and dissolved, a treatment process first removes suspended contaminants by screening, settling, and sometimes filtering. Examples of suspended solids might be grit, food fragments, etc. Dissolved material might be sugar, salt, detergents, milks, and so forth.

Sewage flow is usually at some velocity that prevents suspended solids from settling. The engineer takes this matter seriously when laying out piping systems to avoid pipe-plugging with solids. This is why line size varies depending on anticipated flow rates.

The first area sewage might encounter is a *bar screen*. The purpose is to remove large objects that could block or injure treatment equipment. From here the waste flows to a *grit chamber* of sufficient size to a slow flow rate to settle solids. The waste continues onto a *settling tank* that is another stilling area from which primary sludge will be removed. The material removed from the liquid at these three locations is not stabile and can become very odorous. Together, the bar screen, grit chamber, and settling tank make up *primary basic treatment*.

Wastewater Treatment

Primary sedimentation, or physical separation, will remove 40–60 percent of wastewater suspended solids and about 30–40 percent of biological oxygen demand (BOD). Important to understand is that while the solids are removed from wastewater at this point, they need to be treated in a way that is appropriate to the community and geography of the operation. The solids, termed sludge, provide us with a social challenge. The pollutants have not disappeared, but as we will see, a waste-treatment process fractionates, separates, and then concentrates the residues into sludge. This fraction ultimately needs digestion or disposal in some social, safe, and esthetically-prudent manner.

Secondary Waste Treatment

Two treatment processes may be available to food processors at this stage. The first system deals with a biological method where microorganisms are used to convert organic dissolved solids into biomass. In other words, bacteria use the food supply to reproduce and form more cells. As cell numbers increase they will become floc and, as such, can be separated from the liquid by physical means. The biological process can be either aerobic or anaerobic. At this time most food processors use aerobic methods because of historical success with the process. However, aerobic biological processes produce considerable amounts of sludge that need more treatment. Sometimes sludge-handling may account for 30 percent of the total costs of operating a food plant's waste treatment facility.

Anaerobic treatment produces less sludge but is a much slower process. A plant can process waste streams in aerobic operations in hours or in a day or two, while anaerobic systems may require 20 or more days to treat wastage. If a plant discharges waste volumes of 300,000 gallons or more per day and long retention times are needed, a large land area or huge tank capacity would need to be made available.

Another option now available to separate dissolved organics at the molecular level is the use of membranes to screen residues from wastewater. The technique is physical, quick and effective. As yet, only a few factories are working with the equipment. I suspect more installations will come on-line in the years ahead as additional experience is gained that can be related to reliability and cost.

Secondary Treatment

With secondary treatment aerobic chambers, in which aeration tanks mix food in sewage with bugs that feed on the food, are used. There can be up to 90 percent removal of organics. From the *aeration chamber* sewage flows into a *clarifier,* which is another settling tank from which solids or sludge is removed. The effluent from the clarifier can often be discharged to some watercourse at this point if the stream has sufficient assimilative capacity to naturally purify the remaining impurities in the liquid. Sometimes the effluent is treated with bactericide such as chlorine for safety reasons. Sometimes the firm needs to provide *advanced wastewater treatment* called *tertiary treatment.* These added steps are costly and, for the most part, are used to deal with unconventional waste flows discharged from food plants. This may not be the case with effluents from chemical, steel, and other nonfood operations. For the most part, food wastes are treatable with biological systems, and effluents from the plants can be discharged to land or water receptacles.

Figure 12.3 provides a graphic overview of more conventional systems.

Sludge

Sewage sludge is one of the by-products produced by treating wastewater. We speak of it in terms of being either *raw* or *stabilized.* Raw sludge is produced by various chemical, physical or biological processing methods. The purpose of producing it is to concentrate the solids dispersed in wastewater in such a way as to make them cheaper to treat and easier to handle. Fresh sludge can be pretty odorous and troublesome, and should be looked upon as being a potentially dangerous vehicle for transmitting disease and toxic substances.

Stabilized Material

Stabilized sludge, on the other hand, is raw sludge that has been treated by anaerobic or aerobic digestion, composting, or reaction with chemicals like lime or chlorine. Several processes to reduce odors and destroy pathogenic microorganisms associated with sludge material are available and are in use.

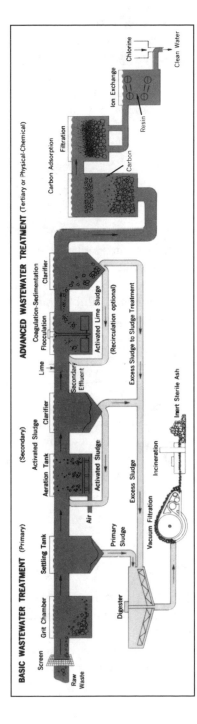

Figure 12.3. Primary, secondary and advanced wastewater treatments are illustrated. At this time most plants are not carrying out a full range of tertiary or advanced wastewater treatment. Treatment plants accelerate organic degradation as might occur naturally.

135

Those of you who have experience with sludge from waste treatment plants know that it looks like heavy fuel oil. Most people are really surprised to learn that the "thick, gooey stuff" is mostly water and contains only 2–5 percent total solids.

Different scientific groups are currently debating the best way of dealing with sludge. There are some who want to use it, and others who want to prevent the use of "processed (stabilized) sludge" for fertilizing or conditioning soil for agricultural purposes.

Sludge Use

The fuss and fury about using sludge may seem odd to some people because both raw and stabilized sewage sludge have been applied to the land in Europe and in the United States for a very long time. Sewage farms have existed in Europe for 300–400 years, and in the United States we have been putting sludge on land for a number of decades.

A number of benefits can be identified by spreading sludge on land. Not in the least is the fact that sludge contains valuable nutrients needed in soil. As large amounts of sludge become more and more available to us by treating wastes, there is a greater need for us to look at it and find new ways to exploit its use.

We can and do fool ourselves, however, by failing to recognize some real and serious problems with sludge. We have to manage sludge in order to maximize its benefits and diminish the unwanted side effects of heavy metals that can be taken up into crops grown in sludge-amended soils. What bothers me, and I think should disturb you, is that we have to be sure the sludge we put on land today, tomorrow, or next year doesn't harm people or the land.

Sea and Land Disposal

Because the disposal of sludge into the oceans was stopped in 1982, the problem of handling mountains of sludge by seacoast communities has become almost catastrophic.

Without the sea for a disposal site, some coastal municipalities will have to look to the land for spreading sludge, and many believe that the nearer the land is to the cities, the better. Some choice agricultural areas still exist near cities, and these locations are being "eyed" by city governments as good sites for the land application of sludge.

The popular press and organic gardening enthusiasts have long suggested that sludge can replace commercial fertilizers. It is not my intent now to either defend or challenge such thoughts. It is fair, however, to say that some plants with sludge problems look to replacing fertilizers with sludge. Imagine how promising sludge with a 3- to 6-percent nitrogen content looks as a resource when a firm produces truckloads of it daily from food-processing wastes. Certainly if the factory operates in a rural area near farmland, the idea of moving sludge onto agricultural land is even more attractive, especially if it can be sold. So what starts out as a method to upgrade liquid effluents leaving food-processing factories on one side of a mass balance equation may easily wind up as a bundle of trouble in a smaller package. Some sludge, depending upon waste source, may contain enough toxic impurities to be extremely troublesome. In such a case, one pollutant has merely been exchanged for another.

Animal Feed

Elsewhere, a newer, more popular outlet for food-plant sludge is used as an animal-feed supplement. What makes the use interesting is that when sludge is dried with spray dryers, roller dryers, or even atmospheric drying alone, it can be incorporated into a wide array of feedstuffs without organoleptically or visually spoiling finished goods. Studies underway at universities and private research centers are assessing the advantages and disadvantages of using sludge with its nutrients as a feed ingredient. In the final analysis, one must discern whether the animal eating the sludge winds up as the "screen" or as the "sink" for toxic materials. If the value of the animal or its by-products is lessened by eating sludge, then this kind of a recycling process complicates rather than improves sludge-handling.

The Quick Fix

Beware of magic solutions that promise to make the operation of waste treatment plants foolproof. Physical methods used to separate suspended solids in wastewater are well-understood and reliable, as long as the engineer took the time to characterize waste flow and composition. It is possible to remove substantial amounts of dissolved solids. Sometimes a system malfunctions; but, more often than not, the real problem originated upstream in the food plant. The processing plant may have put a new process on-line with a waste loss that changed the discharge profile.

The Biological System

As previously stated, wastewater that contains organics is mixed with microorganisms for sufficient time to let the "bugs" decompose waste from complex to simple compounds. At first, one group of organisms outgrows others because food and environment favor the species best able to grow in the "broth." I use the term broth because the nutrients in food waste dissolved in a water carrier are what the bugs need. Some microorganisms find it to their liking to use the by-products of other microorganisms. For their purpose different species of microorganisms have their own diet demand to be able to assimilate a variety of compounds. In other words, the waste of some bugs is the feed of another type of bug. Waste breakdown is efficient when organisms are able to assimilate waste compounds quickly and reproduce rapidly.

If the process is aerobic, much of the food material or waste is oxidized to carbon dioxide and water. The oxidation provides the energy for microbial reproduction while the carbon dioxide and water do not create problems. The new bacterial cells or biomass produce what we term sludge. The sludge is the concentrate or the dissolved solids that can be separated and treated separately. In fact, we create a different problem.

When anaerobic organisms degrade waste, the process is slower and organics are converted to hydrogen sulfide, methane, and other by-products that smell. The familiar term "rotten egg odor" is not without justification. Anaerobic microorganisms obtain reproductive energy by oxidation of complex organics, utilizing compounds other than dissolved oxygen. A textbook definition describes oxidizing agents as electron acceptors, and oxygen is not required for oxidation. In fact, fermentation is the process where organics are used by microorganisms in the absence of oxygen.

Anaerobic organisms are, for the most part, more selective in choice of food than aerobic species, and they produce cells more slowly. Anaerobes can be totally inhibited by small amounts of oxygen, and some need pre-digestion of organics by one group to provide feed for still another. Methane production requires at least two groups of microorganisms in balance to generate gas with any degree of efficiency.

Between aerobic microorganisms and anaerobic species there exist "bugs" termed facultative. Many species exist in either-or environments; but, when given the choice of dissolved oxygen, they metabolize waste more rapidly. We also ought to comment that there are algae organisms that use photosynthesis to process complex waste material into simpler compounds. We need to know that fungi, too, have a role in waste degradation.

In general, the biological process is the more difficult unit operation for waste treatment plant operators to manage. Given food of some broad quality, microorganisms will selectively gravitate towards those species best able to survive and grow in the substrate provided. The organism must, in fact, become acclimated to the wastewater.

Because waste-flows from food plants are rarely uniform in volume rate or composition, they are easily able to upset biological processes. To minimize problems with biological systems use, engineers have devised safety steps to keep treatment plants on-line. First of all, waste flow ought to be combined into some surge tank of appropriate size and configuration to mix waste without letting it go septic. The next step, depending on waste source, may be a method of adjusting pH. This is done by adding acid or alkali to a mixed volume to keep pH under control. You may remember that municipal treatment systems limit influent to pH zones between 6–8.5. Therefore, if you elect to discharge solutions that upset pH, you need adjustment capability. From this point, the liquor flows to the biological treatment phase.

Microorganisms go through a four-growth phase sequence. Most of you recall the biological growth curve that is the bell type. First we see a *lag phase* where growth seems slow because numbers are low and the microorganisms are adjusting to their environment. Next is a *growth phase* where vigorous active multiplication occurs. Food and organism life are in harmony and growth is rapid. As the cells mature we have the *plateau phase*, which signifies decline. Finally there is an *endogenous* or dying-off phase, which brings us to the next point. Some people think one way to keep a system active is to feed the right kind of culture into the waste flow so the preferred organisms can do their job. In my opinion, they fail to recognize that, if the natural selection process of conditions of temperature, food quality, pH, etc. is maintained, the microbial populations of durability will assume their place in the food broth.

Bioaugmentation

I have challenged this concept with dairy industry waste by evaluating the efficiency of the practice. It's a misconception on the part of some people to think the biological treatment of dairy processing wastewaters is a relatively simple task. Historically we are told that many treatment plants operate at less than 70 percent effective 25 percent of the time. A common defect, termed "bulking," occurs where microbial flocs fail to separate and pass into the final effluent as BOD. The problem is often described by the "buzz phrase" filamentous bacteria. This condition overstates the problem. There's a real problem in some parts of the globe where winter weather cools aeration basins to about 41°F (5°C). The ability of microbial flocs to separate and collect is not as good in cold weather as at a warmer temperature. There are other reasons, but they will not be part of the discussion in this general overview.

Working with graduate students, a senior engineer, and technical support staff, two in-depth studies were carried out over a 2-year period. The study to test the value of bioaugmentation, at least for dairy waste, was not positive. Some researchers claim beneficial results with municipal systems, but we have no supporting information to use bioaugmentation (add special cultures to sewage) to make waste treatment systems more efficient.

In the Zall trials there were two separate studies. In Study One, eight bench reactors treated dairy plant wastes. The reactors were seeded with mixed liquor containing microorganisms from two operating dairy plants and operated at steady state. In the second study, four bench reactors were used to treat dairy plant wastes. All reactors were also operated at steady state.

In both studies the reactors were operated in a draw-and-fill mode, and two commercial bioaugmentation cultures (supposed to be suitable for dairy waste) were added to the bench reactors. Considerable manipulations were used to randomize test batches with a goal to obtain reliable statistical information. In final analysis our work showed that the concept of bioaugmentation lacked merit in treating problems of dairy processing wastewater. The best results were obtained by stabilizing those conditions needed to maintain a balanced population of what are termed filamentous and zoogleal microorganisms. As we like to tell treatment plant operators, "If you keep food supply relatively constant to microbial population, temperature, and pH in range attractive to bacteria, the probability is the system will function well."

Looking Ahead

When it comes to thinking about opportunities for better waste management practices in the years ahead, we need to picture what the food factory will look like. Sure, there will be operations that niche the production of exotic foods by unconventional processes, but I suspect the basic need to provide foods manufactured from fruits/vegetables/dairy products/grains, beef/poultry, etc. will remain the same.

While factories will look the same, we can expect to see leaner and meaner workers doing the jobs they do best in fewer but bigger food factories.

What Goes Around Comes Aground

In the past century we have seen food factories change from operations that made single commodity items to multiproduct-producing installations. Seasonally operated fruit and vegetable plants that ran about 150 days/year looked for added product lines to keep staff and equipment busy the year around. Later on, bigger one- or two-product plants seemed better than smaller units, so the food industry constructed huge processing factories in areas unable to handle waste loads previously acceptable from smaller units. Now we see a return to multiproduct food factories, but utilities of steam/water/electricity/waste treatment facilities can be managed to operate more efficiently with little waste. (This is the story being circulated.)

Changes

Areas of change will focus on personnel used to staff the food factory. They will be better-trained, but there will be fewer people. The factory is most likely to be a computer-integrated manufacturing operation. Most firms will divide automated systems into subsections to reduce risk and at the same time remain flexible.

As for change in process technology, basic processes will be the same, but major impact is expected to occur in thermal processing, fractionation, concentration, and dehydration.

Renovation and reclamation of waste streams will be standard practices. For example, stainless steel membranes are expected to be common tools to reuse caustic and acidic cleaning solutions.

Power consumption is to become efficient by the use of strategic energy plans within factories controlled by computerized monitoring systems. These control systems, along with better-designed hardware, will process foodstuff with lower energy use.

When it comes to distribution and warehousing, one only has to look at revolution in moving small packages by Federal Express and like organizations. Automated warehouses will not only be controlled by product, but also by time-temperature indicators as an alternative to open dating or codes.

These changes and others will, no doubt, reduce food wastage in distribution. Total management is apt to become more than just buzzwords in the years ahead.

Environmental Concerns

Expect more of the same in the years ahead when it comes to being limited as to what a food plant can discharge as liquid waste and solid waste. You should be able to address noise pollution as well as air pollution. More emphasis will be given to waste reduction and recycling schemes. In the final analysis the food factory will still be required to treat wastes. More operations may have to deal with tertiary treatment steps to meet demands of regulatory agencies.

Cooperative Systems

Side-by-side food factories of different organizations are likely to become more common in the years ahead because each area is confronted with like restrictions.

Taxes, services, traffic gridlock, distribution network, cost of power, transportation costs, etc. are some of the items that dictate where factories are located. When this happens, side-by-side operations can decide to build and operate a single waste treatment plant to serve two or more organizations. Plant geography dictates environmental constraints and restrictions because the food industry of the future needs to be a friend of the environment.

The author has had firsthand experience with a project where two side-by-side factories owned by different organizations got together to treat waste effluents in a single treatment plant. The site location was Juda, Wisconsin.

What is different about this case is that the waste by-product of one factory would be processed by another. In this example a cheese-producing operation wanted to make more cheese but didn't want to deal with whey. For every pound of cheese made, about 10 pounds of whey is a by-product. The cheese plant had two major waste streams consisting of whey and waste effluent. Processing, cleaning, spillage, and so forth generated plant waste. Whey was waste that was sewage when sent to drains or, in this case, loaded on tank trucks and spread on farms in the general area. The plant that produced whey wanted another company to not only take the full output of whey, but to take care of the sewage effluent generated in the cheese factory.

The firm agreeing to process the whey built a factory alongside the existing cheese plant and proceeded to construct a factory with state-of-the-art technology to convert whey into whey protein concentrate powder and lactose. Figure 12.4 shows the new factory under construction. At the same time, the new plant provided a waste treatment system that would treat effluents from two factories in one common waste treatment facility.

In this case, the plant took its shape in an empty field alongside the existing cheese factory where, within budget constraints, management developed plans and actions needed to build an energy/waste-efficient factory. Figure 12.5 shows how the two plants built a common supply trunk corridor through which whey was pumped.

Figure 12.4. New whey processing plant being constructed alongside an existing cheese factory.

Figure 12.5. Both factories were connected by a corridor through which raw material flowed and clean-in-place piping was installed.

Projection of energy use was such that management felt that, over time, energy costs would rise. The cost of labor would also be greater. Automation, where feasible, should be utilized. Changes in technology were likely to occur; thus, plant flexibility would be desirable.

The factory was designed to be a fractionation, concentration, and dehydration facility. Whey, a dilute raw material, is mostly water at about 94 percent. Logically, the process of first choice was ultrafiltration that fractionates whey into at least two streams with considerable economy of energy. The whey proteins could be segregated at different levels for concentration and then be dehydrated using energy-efficient schemes. The permeate needed to be both concentrated and fractionated too. Here concentration was achieved using a mechanical vapor compression system. Separation of lactose was accomplished by crystallization, followed by dehydration. Here too, a novel energy-efficient dehydration unit was employed.

The Treatment System

Waste treatment is another concentration dehydration operation. Waste liquor of about 99 percent water needs to be fractionated and then concentrated with favorable energy processes. An aerobic wastewater facility using intermittent cycle aeration systems was installed. The system was applied for use in the dairy industry—to my knowledge, for the first time in the United States.

Some innovations that deserve special recognition were incorporated into the new factory. The vapor from evaporation that I called cow water, the use of which was pioneered by this author back in the early fifties, was the major source of water for both hot and cold purposes in the factory. Cow water was not only earmarked for heat exchange purposes, but also as the mainstay for drier protection against fire as the quenching liquid.

Segregation of High BOD Water

Paramount to the planning of the factory was the intent to minimize the use of raw water and to minimize effluent flow both in volume and waste strength. Cow water generated from the evaporator systems was divided into two main streams based on contamination. High BOD water was segregated from low BOD source. Weigand evaporator engineers were able to provide evaporators with designs that allowed the segregation of about 8000 pounds per hour of high BOD water from two separate evaporators with a combined evaporation capacity of more than 90,000 pounds per hour. Segregation was made mainly within the permeate system using a multipass collandria evaporator. This vapor material, with a BOD of about 100 mg/l, was sent directly to the waste treatment plant. The balance of the permeate was sufficiently low enough in BOD quality that it could be discharged directly to a water course without violating the rules of the State's Department of Environmental Control. As for waste treatment, the benefits of the Sequential Batch Reactors (SQBR) as a novel low-energy-demanding system were viewed against conventional aerobic treatment processes. While it would have been useful to think in terms of incorporating selective anaerobic technology alongside the aerobic system, it seems bankers are reluctant to fund projects without proven track records. Perhaps such limitations promote longevity of obsolete processes.

The SQBR operation is relatively simple to operate, and it will handle loads of variable BOD values. The manufacturers of the equipment claimed the process is capable of handling hydraulic peaks up to six times the average design flow. The system requires little maintenance and provides low amounts of sludge, which are relatively stable due to long Solids Retentate Time (SRT).

Final Analysis

Perhaps the future will be a vehicle in which more cooperative programs can take place where wastage from one industry can be the raw material for another.

By accommodation of the separate activities undertaken by the two different companies in this case study, we can conclude that each firm will be able to focus exclusively on what the separate organizations do best. That is, the company making cheese would be able to utilize its capital to grow and to produce more cheese without dealing with its so-called waste streams. The company that manufactures and markets whey residuals will be able to utilize its special talents without having to generate whey and deal with the myriad of cheese-making problems. Perhaps this is the kind of synergistic association that needs more attention.

CHAPTER 13

How to Seek and Gain Help to Solve Waste Problems

I f I were faced with a waste problem in my factory (be it production or treatment), I'd look to people in my organization to see what we know about dealing with the issue.

When and if I did have in-house talent, I'd need to know on what basis I could accept in-house advice. For example, what kind of credentials might the people to whom I looked for help have? Were they trained? Were they experienced? What kind of a track record did the individual have?

I'm reminded of a paper I once gave in Ireland in 1984 at an International Dairy Federation Seminar entitled "Operational And Maintenance Problems In Dairy Effluent Treatment Plants." Excerpts of the address follow:

"Most managers in milk factories now recognize a need for in-plant controls to reduce waste. More people are becoming aware of some savings and the potential economic value of pre-treating waste effluents so as to lessen treatment costs when using public services. It's reasonable to expect that in the future we will see more interest in pretreatment schemes to save operating costs as municipalities charge more money for handling industrial waste flows. It's also true that management in the States is becoming more mature in thinking when faced with a need to operate company-owned waste treatment plants. This change in attitude is due to the fact that high capital investment is needed to build treatment plants, and continued expensive annual operating costs are required to keep waste treatment sites on-line.

"Just as there are differences in waste quality in effluents from similar types of milk factories, we have to remember that we have differences in the way executives view goals to target in short-run and long-run situations. No two plants face the same financial problems or limitations. Managers tend to differ in emphasis they wish to put into waste treatment plants, but it's fair to surmise that most executives would want to spend as little money as possible without creating environmental or legal problems. The preferred, and probably the most economical, effluent treatment works constructed have been those especially tailored to meet specific site situations."

Selection of Sites and Plant Types

Before preparing my paper for the conference I met with people at different dairy plants around the United States and, with their help, assessed some of the more obvious and sometimes even subtle operational and maintenance problems of running effluent treatment plants. Waste from cheese factories was selected as scenario material because it seemed reasonable to me to expect cheese plants to have more waste treatment problems than fluid milk or ice cream plants. Fluid milk dairies are generally located near large cities, where they mostly use public waste treatment systems. Four rural factories with their types of treatment plants were selected.

Plant A—a cottage cheese/whey-drying/skim drying plant
Plant B—a cottage cheese/lipid whey-drying plant
Plant C—an Italian cheese/whey-drying/milk drying/butter plant
Plant D—a cheddar cheese/whey-drying plant

Prior to making site visits to effluent treatment plants the assumption was made that it would be possible to identify more common plant problems and attention could be focused to more serious issues. For the most part, the hypothesis was not true. Some of the queries used to gather information included questions such as:

1. What type of treatment plant is being operated?
2. What is your waste flow per day?
3. What is the waste load in number of pounds of BOD going in and out of the system?

4. Is the waste plant able to treat wastes well enough to meet permit standards? If no, what are the problems with meeting permit discharge limits?
5. Does some particular piece of hardware cause more operational problems than others?
6. Given an opportunity to make changes, what equipment might you like to change?
7. What are the biggest problems with the system being operated?
8. Can you share economic data about costs per day for operating the system?

The type of problems that most people wanted to talk about was not hardware-specific. Contrary to what one might think, personnel running waste treatment plants were generally not willing to fault machinery. Perhaps the plants I visited were not old enough. For the most part, data were collected from relatively new plants (5 to 10 years old).

As you might suspect, the product processing areas of plants, together with area supervisors, were blamed for waste treatment plant malfunction due to creating overload conditions. One exception to such comments was a plant where the factory superintendent took charge of the day-to-day control and instructed treatment plant personnel as to how the sewage plant was operated. Sure enough, this treatment plant looked well and operated to appropriate standards.

More than one complaint was heard about the advice provided by consultants who were unable to provide explicit instructions on how best to run waste plants. Managers tend to feel they may not be getting real dollar value for money spent. Of course, there are more than one or two sides to these kinds of complaints.

Selective complaints registered by operators of the four different effluent treatment plants were as follows:

Plant A noted that electronic equipment was too complex and could not be repaired in the field.

Plant B reported that aerators were unable to supply sufficient oxygen to the system. They noted they had poor consultants. The comment was made that the agitators had to be run regardless of oxygen content because the system needed to be agitated.

Plant C seemed to have a complaint that the sludge-drying apparatus had poor belt speed control and the system operated with an undersized clarifier.

Plant D complained that composite sampling equipment, clarifiers, and the air distribution system were poorly designed. The site operator was vocal in complaining that consultants were poor and provided few explicit instructions on how best to operate a waste treatment system.

Consultants

The old adage that "free advice isn't worth much" or "you get what you pay for" is not always true. Just as you need to pick a good physician to remedy personal illness, consultants fill the same bill.

The same value judgment you make in turning to inside talent to solve problems needs to be applied when selecting consultants.

(a) What experience does the consultant bring to the problem-solving exercise?
(b) What credentials does the consultant possess (organizational backup is also important)?
(c) Check the consultant's track record.

There are other issues you might wish to look at, but the issues above make a good beginning.

Consultants market two principal items: "know-how" and time. If a firm waits to get the most for its money it needs to do its homework before involving consultants. For example, you might want to check with consultant X as to what kind of on-site information needs to be made available or collected prior to meeting the advisor. Ask if you can send consultant X information before he or she invests a lot of time on a proposed project.

Large and Small Organizations

Large food organizations with multiplants usually staff professional engineers to take care of worldwide operations. Even with in-house talent, the large company will seek outside specialists or consultants to help with complex or totally new issues. This is all well and good, but what's a small organization supposed to do when it needs professional expertise?

In my case, with relatively small operations, I looked for outside expertise as follows:

I contracted with a well-known P.E. (professional engineer) who was moonlighting from his regular job as Director of Facility Engineering at a nearby location. He served as president of the regional engineering society, and I looked into his academic training before retaining his services. The raw truth was that my consultant was locked into his full-time job by "golden handcuffs." For some, the term might be unfamiliar. What it means is that, through years of service and perks, the employee gives up a great deal by leaving or changing employment. I guess tenured professors suffer or enjoy a similar affliction.

A deal was struck that, for a set fee, the consultant visited my site three Saturdays a month. He would tour plant operations, review power consumption figures, look at waste treatment facilities, and take a cursory view of machinery. He would hold face-to-face meetings with me post his site inspection and follow up our discussion with a written report.

The benefits from the arrangement were remarkable. First of all, regular review and inspections turn up unexpected problems. Outsiders see issues local people overlook or accept as being normal.

Over time I learned to save problems and questions all week prior to our meeting for discussion with my consultant. Over many years the association proved to be beneficial to both of us because he taught me engineering and I educated him on the dairy industry. When special projects needed to be addressed, my consultant was smart enough to bring along other professionals with expertise.

Sooner than you might think, I was able to draw on drafting, engineering, equipment layout, purchase specifications, etc. Both parties profited, and I think a good deal took place.

The message is that small companies can hire outside talent for special problems and then separate the expensive talent when not needed. Larger organizations usually retain an array of specialists full-time and apply other rationale for such services.

Suppliers

The days when sales people came in with a new joke, a gimmick or a small gift to assist (they might have thought) sales are long gone. We need to remember the past to keep it out of the future. A sales technical individual is expected to be a specialist. The sales representative is a valuable talent that can give a potential buyer extensive information about the equipment being presented: market information, track record, etc. Major suppliers can have an array of their own in-house talent that might be useful to an operation thinking about purchasing material, machinery, or whatnot. The supplier can be part of a network source of information and needs to be brought into your own operation where a "good fit" is advantageous.

Government Agencies

When it comes to getting useful information, the food manager needs to exploit state and federal agencies for "freebies." I have to smile when I use the term freebies because, when I used the term in classroom lectures, foreign students needed additional term definition. The United States and Canada are prime examples where it seems no expense is spared to produce and send off bulletins, pamphlets, books, videos, computer disks, etc. to deal with environmental issues.

A manager who fails to plug his organization into agencies with the variety of "freebies" that can be funneled into his organization can be likened to an individual who fails to bend over to pick up a bundle of money left on the sidewalk.

Besides information that can be acquired via mail, some governmental agencies operate technical service departments that send out specialists to businesses solely to assist in technology transfer activities. Sometimes the governmental agencies operate in concert with colleges and universities in technology transfer programs. What helps alert a manager to these programs is the need to be put on different mailing lists. It's both a responsibility and opportunity for plant managers to become part of ongoing educational programs.

Colleges, Universities, and 2-Year Programs

As a longtime extension program worker in education, I've seen some strange practices and logic used by food factory managers. For example, "Do not become involved with university specialists. . . . They really have little to contribute to our business, and they go away with our factory secrets." Can you imagine how foolish this attitude must seem to specialists such as extension agents? By extension personnel, agents, or whatever terms some parts of the world use to identify these individuals, you need to know that state colleges (and universities) hire teachers to extend educational programs off-campus. In fact, extension specialists and professors have been a "secret weapon" used in the

United States to move agriculture from rural America to a world-class, technically-proficient society. Were it not for extension personnel input, we might not be in the forefront with animal breeding programs, soil technology, crop harvesting, etc. etc.

It's almost a laugh to know that, for the most part, industry secrets are secret only to those lazy enough not to use libraries or subscribe to an array of technical journals.

Having made these claims, which may offend some of my readers, over the years the only secrets I've seen are those that people didn't know were not secrets at all.

The down side of Extension Technology Transfer follows the adage that squeaky wheels get the grease. I think that most of my extension efforts (my responsibilities at Cornell were in Teaching, Research, and Extension, where most faculty members are encouraged to select two of the three areas cited) focused on but a few clients. Most industry managers seemed to be of the opinion that if one does not pay for advice it's probably worth little. On the other hand, some astute organizations go out of their way to cultivate university people to work with them in problem-solving and to innovate research programs. You need to remember that many centers of education have more talent and facilities to carry out research (applied and theoretical) than industry food factories. Of course, they are exceptions, but what has been written is factual.

The fear of sharing information can be troublesome if we agree that public educational systems are not supposed to do secret work that cannot be published.

There are changes with this concept; and more private arrangements are being negotiated between the private sector and universities, especially in genetic engineering and biotechnical areas.

We see little interaction between industry staff with 2-year colleges or technical programs carried on by trade-school educators. In New York State there is a variety of 2-year colleges located in some counties. In addition, we have BOCES (Board of Co-op Educational Services) applied training programs carried out during days and evenings in educational centers. Subject matter is varied and taught by experienced teachers. Of late, computer graphics and nutritional studies seem to be favored subjects.

I do not mean to criticize the "think tank centers for Professional Advancement" operated by a part-time faculty of professionals who hold the 2- to 3-day regional training courses. These are good, expensive, and short-lived. Often, firms send the wrong people to the seminars. What makes some of the programs inefficient is that, too often, technical transfer is not followed up by participants once back to their home factories. In other words, I fail to see a big bang for the buck. There could be a considerable variety of low-cost training programs made available to more industry workers if management recognized that better-informed and -trained workers could do more assignments in our factories. We can grow as the work force grows in technical expertise.

Trade Organizations

We in the food industry are fortunate to have had a cadre of selfless individuals put together organizations like the Old National Canner, Milk Industry Foundation, International Food Technology (IFT), and different health/industry cooperative interest groups. There is much to be pleased with, and technical progress was certainly enhanced by the efforts of these and other organizations.

However, it is not enough that we train and educate our professionals and keep them abreast of new information. For the most part, trade organizations have not trained their "big guns" on blue-collar line workers. In my opinion, we have a dismal record in this area. In Europe and elsewhere in the world we find ongoing healthy training programs to upgrade the skills of the common labor force. Some trades send workers to factory schools operated by the trade organization, which offer full pay and other perks and make the employee a more valuable asset. Trade schools are sometimes small factories with industry-scale equipment. Workers are taught chemistry, physics, computer process control, etc. I've had an opportunity to visit and work with people running these programs, and I am enthusiastic about the payback on the investment. It is not enough to focus just on top management; the entire work force is important. I'm apt to remind the reader of Napoleon's sage comment, "Every soldier carries a marshal's baton in his knapsack." So it may be with many individuals in our organizations.

Present and Future

In waste management, the easy road is to move along with the *as is* mentality. For some, the very idea of initiating a new program is painful, fearful, and unwelcome. The trick to move ahead is in the *to be* mode, which takes leadership and vision. Managers or non-supervisors have to feel that there may be better, easier, more productive, more profitable, and superior methods of operating their factory. *What if* we made an investment of providing the food factory with on-site study centers where the work force can thumb through trade journals, textbooks, trade papers, and the like? What is being suggested is that every factory operates a small lending library dedicated to supplying information appropriate to the commodity being processed. What if there were a couple of computers and a Xerox machine in the education center? What if the person in charge of the small library was able to reach out via electronic mail for abstracts? What if managers assumed that a more informed worker might be a better worker, and so forth? In this area, we have much to learn. We do ourselves a disservice by thinking that only research people need the kind of informational services being suggested for the entire work force. There are available subscriber information services. For a fee, the service provider allows a company to conduct literature review from its own office via telephone and to receive printout data on writer- or faxlike machinery. The technique is to dial the agency and list or cite subject matter, type of data bank, and years to be covered, then list 3–4 keywords. Within a few minutes a printout containing information spews from the machine. In other words, there's no real excuse for a manager to be ignorant of what was, is, and may be available for use in the field.

Libraries and the Internet

Unless you do business in an area where a college or university is located, it may not be easy to retrieve food science textbooks or appropriate trade journals. Having made this statement, it ought to be evident that within most states you are able to ask your local library to borrow the texts you seek from interlibrary lending practices. It may take a couple of weeks to get the material, but if you are able to wait for the information, such services may be available.

You can become a frequent visitor to your nearest college food science department. Often, the department maintains a limited library where you can retrieve information and, still better, talk face-to-face with people who may assist you. You can then make arrangements with the college central library for your organization to use the facility. In other words, it's up to you to make the effort to utilize the inventory of information you might wish to acquire. As I previously noted, "freebies" are all around you as a smorgasbord or banquet just for the asking. You can be no farther away from useful information than your computer and websites.

Local Wastewater Treatment Plants

Municipalities have updated or constructed treatment plants with state-of-the art technologies. These facilities are being staffed with professionals who are knowledgeable in the art. You should not miss the opportunity to visit the sites and become friendly with the staff. Sometimes help with your very own problems may well be just down the road from your food factory. An up-to-date waste treatment plant, equipped with laboratories, can be useful to you and your own people with technical transfer information.

Select Informational Sources

Managers and supervisors can seek information from many sources. In general, we can group publications in several categories:

Food Processing Industry Periodicals
Water and Waste Periodicals
Water and Waste Books by Commodity, Management, By-Product Recovery, and Environmental Law
Environmental Protection Agency Publications
Individual State Wastewater and Solid-Wastes Control Agencies
University Extension Publications
Select Government Websites

Each year, up-to-date information is published about governmental agencies. The reader will find organizational charts as well as names and addresses with telephone numbers of key people in different administrative slots. One journal that comes to mind is *Food Processing*. You may also contact your trade organization for a Who's Who list of personnel and agencies you might contact for assistance. Remember, these people are not Internal Revenue Tax agencies from which many individuals shrink or want to forget. These agencies were created not only to solve problems, but often to assist the very industry they are supposed to regulate.

Industrial Waste Conferences

Over the years we have had the Purdue Industrial Waste Conference that attracts world-class engineers who relate case study information and present research papers. There are conference proceedings that are useful to firms looking for answers to select problems. Georgia Tech Research Institute provides proceedings of the Annual Food Industry Environmental Conference. This is a very useful conference for food engineering staff. The EPA holds technology-transfer meetings that will assist food factory supervisors. Many universities publish information about ongoing food waste research programs. These bulletins need review by today's food industry managers. The same holds true for non-USA firms. The Europeans, Australians, and, of course, other world centers provide select technology transfer material.

This chapter only touches the fringe of the available information that's available to people in the food industry. I remember a statement one of my old professors in the University of Massachusetts made back in 1946–47: "You are only as smart as your ability to get information quickly from your library."

By this the instructor meant that you cannot clutter your mind with a lot of individual facts (this is good too) as opposed to being able to remember where to reach out quickly to informational sources with a wide range of facts.

With the rapidity of technological changes now taking place in the world, and with an array of quick information systems being available, the more progressive managers need to plug into networks that keep them sharp. None of us wants to become redundant because we fell behind. Your ego needs to be primed to keep you fresh.

CHAPTER 14

Self-Test

Can You Pass This Test?

Just as supervisors are expected to understand each unit process in the food factory, it seems to me that the individual ought to know about waste management and the factory's sewage treatment facility. Some years ago I introduced a course in waste management to students in Cornell's Department of Food Science, and over time it became an area of concentration. From Day One the message was given that the material being offered was not supposed to train the student to become a civil engineer. The goal was to teach students that, as food scientists, they had an important role in making sure the food plant was environmentally friendly. Spin-off advantages to students were that they could not only become responsible for developing products, but also be able to consider wastage both in amount and type when putting a new product on-line.

The following test represents the minimum amount of information I think a food science person ought to know about waste treatment technology. If you are not able to earn a 90-percent grade by taking the student exam provided, you need to fill in the training gaps you have in this area. I've included a set of answers to the test in the appendix. The test was given to seniors and a couple of graduate students in 1991. Question IX can be skipped or changed to conform to your plant and your town.

TEST

Numbers in parentheses before questions indicate the maximum number of points to be earned. A five-point bonus question brings the total to 105 points.

(5)   I. This is a warm-up question:

Complete the blanks below with the most appropriate word or
words to best fit the statements. Do not use a term more than once.

Septic tank	Aeration	Sludge
Lagoon	Population equivalent	Dissolved oxygen
Volatile matter	BOD	Oxidation pond
COD	Sewage	Total solids

1. Bubbling air through a liquid is called _____.

2. _____ is an indirect measurement of
   biologically degradable material.

3. BOD values can be correlated with _____
   values obtained from the same waste.

4. The BOD test basically measures the _____
   in diluted sewage samples at the end of 5 days and compares
   it with the result of the dilution water blank values.

5. A sewage pond is often called a _____.

6. Waste loads are frequently compared in terms of _____
   _____.

7. Water is frequently called _____ after it is fouled.

8. The bottom of a lagoon will often contain _____.

9. _____ is a rough equivalent to the organic fraction in total solids.

10. An aerobic lagoon is synonymous with _____.

(10)   II. A food plant discharges 500,000 gallons of liquid waste per day.
        The waste has been analyzed for some basic wastewater character-
        istics, which are:

BOD# = 200 mg/l; pH = 7.0; TS = 100; COD = 400 mg/l; SS =
75; Hexane soluble = 10 ppm; Imhoff cone settleable solids = 20 ml

1. How many pounds of BOD/day does the food plant discharge?
   (Show calculations: Assume weight/gal = 10 pounds)

2. What would the population equivalent of this waste load be?

3. What is the BOD/COD ratio?

(5)   III. Waste treatment methods are usually divided into PRIMARY, SECONDARY, and TERTIARY systems.

      1. What is meant by primary?

      2. What is meant by secondary?

      3. What is meant by tertiary?

(10)   IV. True or False

      1. _____ A grit chamber is a tertiary treatment mechanism.

      2. _____ A trickling filter is used to regulate flow rates.

      3. _____ A bar screen is a laboratory analysis for alcohol.

      4. _____ Extended aeration is similar to anaerobic digestion.

      5. _____ Biological aeration systems operate better when BOD:N:P ratios are maintained at 100:5:1 amounts.

      6. _____ Sludge is always variable and never comparable.

      7. _____ Flocculation media are used to purify water so it tastes good.

      8. _____ Methane gas is generated from aerobic digesters.

      9. _____ Waste in sanitary sewage lines can be diluted in stormy weather by infiltration from groundwater.

     10. _____ Physical systems treat waste in minutes or hours, while biological systems can take hours or even days.

(10)   V. Your instructor has discussed the merits of "marrying" wastes to existing edible products. Of course, you have to do a "mass balance" of the waste flows in a food plant in order to know what might be feasible to initiate a salvage scheme. Assume the plant in question produces chicken noodle soup and packages the product in 1-gallon plastic pails for university kitchens. Dressed whole chicken packed in ice is supplied by a nearby poultry processing plant. Noodles are purchased in bulk from Mueller's Noodle Corporation in 50-pound bags. Describe a plant waste survey. List some advantages to be gained by a firm in conducting this kind of a survey.

(10)   VI.   True or False

1. _____ Grab sampling is the same as proportional sampling.

2. _____ Composite samples represent a weighted-average sampling system.

3. _____ A 2-inch reading on a 60 degree V-notch weir or a 90 degree V-notch weir provides the same flow volume reading.

4. _____ Parshall flumes, as the name implies, measure partial flows.

5. _____ Both V-notch flumes and Parshall flumes are self-cleaning.

6. _____ Some states use water laws that incorporate riparian, prescription, and something in between systems.

7. _____ Water laws are relatively new and are now necessary because of increased population.

8. _____ Riparian water law works best where water is sparse, such as in arid lands.

9. _____ To be entitled to prescription rights, one does not have to own land adjacent to the water supply.

10. _____ Public Law 92-500 deals with water pollution issues.

(5)   VII.   Waste color intensity may be an indicator of sewage strength. Your instructor presented case studies that treated egg breaking plant waste and milk plant wastes. In which waste stream might you use color as an index of pollution strength?

(5)   VIII.   Draw or list the different steps in your own factory's waste treatment plant. You might wish to divide the system into sections.

(5)   IX.   If you visited a nearby municipal waste treatment plant, draw or list the different steps in the waste treatment plant.

Note: Use another piece of paper if you think it necessary.

(10)   X. Discuss where you think a trained food science individual fits into a waste management scheme in the two situations listed under Parts 1 and 2 of this question.

   1. Designing a waste treatment plant.

   2. Abatement of waste by improving in-house processing technology.

(10)   XI. Define or explain the following:

   1. COD

   2. BOD

   3. Total solids

   4. Suspended solids

   5. Total volatile solids

   6. Weirs

   7. Flumes

   8. Clarifier

   9. Activated sludge

   10. Sludge

(5 bonus)

   XII. More definitions and explanations.

   11. Digester

   12. Flocculation agent

   13. Hexane soluble

   14. Imhoff cone

   15. Screening

(5) XIII.

   1. Sewage sludge is the residual solid-liquid slurry generated by the treatment of wastewater. Why does it need to be dewatered before it's sent to landfill sites?

2. What might sludge from a food plant contain?

3. What is stabilized sludge?

4. Sludge can be treated aerobically/anaerobically and some-times composted too. Cite two additional methods to dewater and stabilize sludge material.

(5) XIV. Summary Question:

This diagram of a wastewater treatment system lacks names for the different components listed. Name items 1 through 11 and tell me what each step will accomplish. (Attach an appropriate figure.)

Name	Duty
1.	1.
2.	2.
3.	3.
4.	4.
5.	5.
6.	6.
7.	7.
8.	8.
9.	9.
10.	10.
11.	11.

(5)   XV. List items (i.e., 6, 3, 10) that create a Primary System.

List items that create the Secondary System.

List items that create the Tertiary System.

Appendix: Answers To Examination

Numbers in parentheses before questions indicate the maximum number of points to be earned.

(10) I. Complete the blanks below with the most appropriate word or words to best fit the statements. Do not use a term more than once.

Septic tank	Aeration	Sludge
Lagoon	Population equivalent	Dissolved oxygen
Volatile matter	BOD	Oxidation pond
COD	Sewage	Total solids

1. Bubbling air through a liquid is called __**aeration**__.

2. __**BOD**__ is an indirect measurement of biologically degradable material.

3. BOD values can be correlated with __**COD**__ values obtained from the same waste.

4. The BOD test basically measures the __**dissolved oxygen**__ in diluted sewage samples at the end of 5 days and compares it with the result of the dilution water blank values.

5. A sewage pond is often called a __**lagoon**__.

6. Waste loads are frequently compared in terms of **population equivalent**.

7. Water is frequently called __**sewage**__ after it is fouled.

8. The bottom of a lagoon will often contain __**sludge**__.

9. __**Volatile matter**__ is a rough equivalent to the organic fraction in total solids.

10. An aerobic lagoon is synonymous with __**oxidation pond**__.

(10)   II. A food plant discharges 500,000 gallons of liquid waste per day. The waste has been analyzed for some basic wastewater characteristics, which are:

BOD# = 200 mg/l; pH = 7.0; TS = 100; COD = 400 mg/l; SS = 75; Hexane soluble = 10 ppm; Imhoff cone settleable solids = 20 ml

1. How many pounds of BOD/day does the food plant discharge? (Show calculations: Assume weight/gal = 10 pounds)

$$\textbf{BOD\#} = \frac{\textbf{Number of gal} \times \textbf{\#/gal} = \textbf{conc}}{\textbf{1,000,000}}$$

$$\textbf{BOD\#} = \frac{\textbf{500,000} \times \textbf{8.3\#/gal} \times \textbf{200}}{\textbf{1,000,000}} = \textbf{830\#}$$

2. What would the population equivalent of this waste load be?

$$\frac{830}{.17} = \frac{4882}{.2} \textbf{ (accep) 830 = 4150}$$

3. What is the BOD/COD ratio?

$$\frac{\textbf{BOD}}{\textbf{COD}} = \frac{\textbf{200}}{\textbf{400}} = \textbf{.50}$$

(5)   III. Waste treatment methods are usually divided into PRIMARY, SECONDARY, and TERTIARY systems.

1. What is meant by primary? **First major treatment is sedimentation/physical; that is, physical separation barriers, grit chambers**

2. What is meant by secondary? **Treatment of wastes by biological methods after biological degradation primary methods of sedimentation**

3. What is meant by tertiary? **Beyond normal conventional second-day treatment for the purpose of advanced water treatment of increasing water reuse; that is, potential**

(10)   IV. True or False

1.   __F__ A grit chamber is a tertiary treatment mechanism.

2.   __F__ A trickling filter is used to regulate flow rates.

3. __F__ A bar screen is a laboratory analysis for alcohol.

4. __F__ Extended aeration is similar to anaerobic digestion.

5. __T__ Biological aeration systems operate better when BOD:N:P ratios are maintained at 100:5:1 amounts.

6. __T__ Sludge is always variable and never comparable.

7. __F__ Flocculation media are used to purify water so it tastes good.

8. __F__ Methane gas is generated from aerobic digesters.

9. __T__ Waste in sanitary sewage lines can be diluted in stormy weather by infiltration from groundwater.

10. __T__ Physical systems treat waste in minutes or hours, while biological systems can take hours or even days.

(5) V. Your instructor has discussed the merits of "marrying" wastes to existing edible products. Of course, you have to do a "mass balance" of the waste flows in a food plant in order to know what might be feasible to initiate a salvage scheme. Assume the plant in question produces chicken noodle soup and packages the product in 1-gallon plastic pails for university kitchens. Dressed whole chicken packed in ice is supplied by a nearby poultry processing plant. Noodles are purchased in bulk from Mueller's Noodle Corporation in 50-pound bags. Describe a plant waste survey. List some advantages to be gained by a firm in conducting this kind of a survey. (Such as but not limited to below)

Advantages
Provide data for a water management program
Detection of losses
Water conservation and recycling
Aid in designing a treatment system
To help prevent embarrassment
Determine if form complies with regulation
Survey sets out to follow the flow of pollutants or waste through a facility. What's needed is determination of the flow amount and waste concentration of waste at different unit operations as well as the total flow, etc. through.
Prepare flow sheet

Construct a sewer map
Locate and install or use sampling stations
Coordinate information or tasks with production staff so as not
   to be considered a nuisance
Establish analytical considerations

(10)   VI.   True or False

1.   __F__   Grab sampling is the same as proportional sampling.

2.   __T__   Composite samples represent a weighted-average sampling system.

3.   __F__   A 2-inch reading on a 60 degree V-notch weir or a 90 degree V-notch weir provides the same flow volume reading.

4.   __F__   Parshall flumes, as the name implies, measure partial flows.

5.   __F__   Both V-notch flumes and Parshall flumes are self-cleaning.

6.   __T__   Some states use water laws that incorporate riparian, prescription, and something in between systems.

7.   __F__   Water laws are relatively new and are now necessary because of increased population.

8.   __F__   Riparian water law works best where water is sparse, such as in arid lands.

*9.   __T__   To be entitled to prescription rights, one does not have to own land adjacent to the water supply.

10.   __F__   Public Law 92-500 deals with water pollution issues.

*Note: Prescriptive rights are acquired by diverting and putting to use, for a period specified by stating, water to which other parties may or may not have had prior claim. (Established by water rights of individual states.)

(5)   VII.   Waste color intensity may be an indicator of sewage strength. Your instructor presented case studies that treated egg breaking plant

waste and milk plant wastes. In which waste stream might you use color as an index of pollution strength?

Milk (No)
Egg (Yes)

(5) VIII. Draw or list the different steps in your own factory's waste treatment plant. You might wish to divide the system into sections.

Primary	Secondary	Tertiary
grit	Biological	Ion exchange
sedimentation	Towers	or
	Filter	RO
	Clarifier	Flox chem
	Sludge	NH₃ stripping
	etc.	

(10) IX. If you visited a nearby municipal waste treatment plant, draw or list the different steps in the waste treatment plant.

(5) X. Discuss where you think a trained food science individual fits into a waste management scheme in the two situations listed under Parts 1 and 2 of this question.

   1. Designing a waste treatment plant.

     a. Define unit operations—characterize process streams and develop mass balance of actual unused goods yields versus pounds of raw materials processed.

     b. Suggest product and process modifications that tend to waste less food and use less of utilities like stream, water, etc.

     c. Work with design engineers and supply those data that engineers need to define a waste load.

     d. Try not to let an engineer overdesign a waste plant because cement and steel systems usually mean one will not improve.

   2. Abatement of waste by improving in-house processing technology.

     a. In-plant water conservation

   b. **Recycling systems**
      Flume water
      Retort cooling water
      Brine recycling
      Detergent recycling
      Renovation and reclamation of waste

   c. **Implant modifications to prevent waste**
      Screening
      Dry-cleaning floors
      High-pressure, low-volume washup
      Gels
      Dry caustic peelings
      Steam blanching
      UF-RO Systems
      New fish deboning methods
      Closed loop

(15)   XI. Define or explain the following:

   1. COD—**(See Glossary)**

   2. BOD—**(See Glossary)**

   3. Total solids—**Solids in a sample left when evaporation occurs at temperatures of 105°**

   4. Suspended solids—**Undissolved substances retained on a 0.45 micron filter.**

   5. Total volatile solids—**Those parts of the solids last by ignition in a furnace at 500–550°.**

   6. Weirs—**(See Glossary)**

   7. Flumes—**Self cleaning dam structure used to measure flow.**

   8. Clarifier—**Settling device or tank to separate suspended solids.**

   9. Activated sludge—**(See Glossary)**

   10. Sludge—**A by-product of wastewater treatment accumulated solids concentrated during treatment.**

(5 bonus)

XII. 11. Digester—**A vessel or apparatus where organic material is degraded.**

12. Flocculation agent—**Substance that can neutralize or change particles so they might coagulate so as to move readily or settle.**

13. Hexane soluble—**Solid extraction of fats/oil/grease by use of some organic solvent.**

14. Imhoff cone—**Laboratory device used to test wastewater settling characteristics.**

15. Screening—**Mechanical method to separate undissolved solids of some size from a water carrier.**

(5) XIII. Sludge

1. **Sludge is mostly water at about 99–95 percent. By dewatering it's possible to save hauling expenses. Dewatering also makes it easier to deal with sludge at disposal sites as it becomes evident.**

2. **Because sludge is a by-product of the food we eat, it contains many of the same elements such as nitrogen, phosphorous, potassium, iron, calcium, sodium and other elements.**

3. **Stabilized sludge is a relatively odor-free by-product of solids extracted from treating wastewater that contains few pathogenic organisms that can be disposed of with minimum hazard.**

4. **Sludge can be dewatered using sand filters, centrifugation, filter press, and vacuum filtration among different systems.**

(5)   XIV. Summary: components and function of each in a wastewater treat-
              ment system

Name	Duty
1. Screen	1. Safeguards equipment
2. Grit Chamber	2. Removes solids
3. Settling Tank	3. Removes solids
4. Aeration Tank	4. Biological reactor where microorganisms convert dissolved organics to cells
5. Clarifier	5. Settling basin
6. Flocculation Chamber	6. Coagulation chamber where chemicals are added to congeal solids
7. Clarifier	7. Settling
8. Carbon Adsorption Process	8. Tertiary treatment step
9. Ion Exchange	9. Tertiary treatment step
10. Chlorination Unit	10. Render effluent free of pathogenic bacteria
11. Vacuum Filtration	11. Solids separation unit

(5)   XV. Primary Systems = **1, 2, 3**

            Secondary Systems = **4, 5**

            Tertiary System = **6, 7, 8, 9**

Glossary

Definition of Terms Used in Water and Waste Management

acre-foot—(1) A volume of water 1 ft. deep and 1 acre in area, or 43,560 cu. ft. (2) A 43,560-cu.-ft. volume of trickling filter medium.

activated sludge—Sludge floc produced in raw or settled wastewater by the growth of zoogleal bacteria and other organisms in the presence of dissolved oxygen and accumulated in sufficient concentration by returning floc previously formed.

advanced waste treatment—A term including any treatment process applied for renovation of wastewater that goes beyond the usual 90–99 percent oxygen demand and organic solids removal of secondary treatment. May include nitrogen, phosphorous, other minerals, taste, odor, color, and turbidity removal by a variety of conventional and special processes as required to renovate wastewater for intended reuse.

aerated pond—A natural or artificial wastewater treatment pond in which mechanical or diffused-air aeration is used to supplement the oxygen supply.

aeration—The bringing about of intimate contact between air and a liquid by one of more of the following methods: (a) spraying the liquid in the air, (b) bubbling air through the liquid, (c) agitating the liquid to promote surface absorption of air.

aerobic—(1) A condition characterized by an excess of dissolved oxygen in the aquatic environment. (2) Living or taking place only in the presence of molecular oxygen.

aerobic bacteria—Bacteria which require the presence of free (dissolved or molecular) oxygen for their metabolic processes. Oxygen in chemical combination will not support aerobic organisms.

173

assimilative capacity—The capacity of a natural body of water to receive: (a) wastewaters, without deleterious effects; (b) toxic materials, without damage to aquatic life or to humans who consume the water; (c) BOD, within prescribed dissolved oxygen limits.

biochemical oxygen demand—A standard test used in assessing wastewater strength. (see BOD.)

biodegradable detergent—One that decomposes quickly as a result of the action of organisms, eliminating foam in wastewater. Biodegradable is defined as having at least 90 percent surfactant reduction, or as having surfactant concentration no higher than 0.5 mg/l.

biological filtration—The process of passing a liquid through the medium of a biological filter, thus permitting contact with attached zoogleal films that adsorb and absorb fine suspended, colloidal, and dissolved solids and release end products of biochemical action.

BOD—(1) Abbreviation for biochemical oxygen demand. The quantity of oxygen used in the biochemical oxidation of organic matter in a specified time, at a specified temperature and under specified conditions. (2) A standard test used in assessing wastewater strength.

BOD load—The BOD content, usually expressed in pounds per unit of time, of wastewater passing into a waste treatment system or to a body of water.

BOD:N:P ratio—The ratio, based upon analysis of wastewater passing into a waste treatment system, of the BOD to total nitrogen to total phosphorus contained in the waste stream. To assure a nutrient balance within a biological treatment system, a ratio of 100:5:1 is generally recommended.

breakpoint chlorination—Addition of chlorine to water or wastewater until the chlorine demand has been satisfied and further additions result in a residual that is directly proportional to the amount added beyond the breakpoint.

chelating agent—A chemical or complex which causes an ion, usually a metal, to be joined in the same molecule by both ordinary and coordinate valence forces.

chlorine-contact chamber—In a waste treatment plant, a chamber in which effluent is disinfected with chlorine before it is discharged to the receiving waters.

coagulant—A compound responsible for coagulation; a floc-forming agent.

COD—Symbol for chemical oxygen demand.

common sewer—A sewer in which all owners of abutting properties have equal rights.

composite wastewater sample—A combination of individual samples of water or wastewater taken at selected intervals, generally hourly for some specified period, to minimize the effect of the variability of the individual sample. Individual samples may have equal volume or may be roughly proportionate to the flow at time of sampling.

degradation of organic material—Reduction of the complexity of a chemical compound by biological action.

deionized water—Water that has been treated by ion-exchange resins to remove cations and anions present in the form of dissolved salts.

denitrification—(1) The conversion of oxidized nitrogen (nitrate and nitrite-N) to nitrogen gas by contact with septic wastewater solids or other reducing chemicals. (2) A reduction process with respect to oxidized nitrogen.

deoxygenation—The depletion of the dissolved oxygen in a liquid either under natural conditions associated with the biochemical oxidation of organic matter present or by addition of chemical reducing agents.

digested sludge—Sludge digested under either aerobic or anaerobic conditions until the volatile content has been reduced to the point at which the solids are relatively nonputrescible and inoffensive.

dissolved oxygen (DO)—The oxygen dissolved in water, wastewater, or other liquid, usually expressed in milligrams per liter, parts per million, or percent of saturation. Abbreviated DO. In unpolluted water, oxygen is usually present in amounts up to 10 ppm. Adequate dissolved oxygen is necessary for the life of fish and other aquatic organisms. About 3–5 ppm is the lowest limit for support of fish life over a long period of time.

effluent—(1)A liquid which flows out of a containing space. (2) Wastewater or other liquid, partially or completely treated or in its natural state, flowing out of a reservoir, basin, treatment plant, or part thereof. (3) An outflowing branch of a main stream or lake.

eutrophication—The normally slow aging process by which a lake evolves into marsh and ultimately becomes completely filled with detritus and disappears. In the course of this process, the lake becomes overly rich in dissolved nutrients.

facultative bacteria—Bacteria that can adapt themselves to growth and metabolism under aerobic or anaerobic conditions. Many organisms of interest in wastewater stabilization are among this group.

field capacity—(1) The quantity of water held in a soil by capillary action after gravitational water is removed. It is the moisture content of a soil, expressed as a percentage of the oven-dry weight, after the gravitational or free water has been allowed to drain, usually for 2 to 3 days. (2) The field moisture content 2 or 3 days after a soaking rain. (3) The moisture content to which each layer of soil must be raised before water can drain through it.

five-day BOD—That part of oxygen demand associated with biochemical oxidation of carbonaceous, as distinct from nitrogenous, material. It is determined by allowing biochemical oxidation to proceed, under conditions specified in Standard Methods, for 5 days. See first-stage biochemical oxygen demand: BOD—(1).

floc—Gelatinous or amorphous solids formed by chemical, biological, or physical agglomeration of fine materials into large masses that are more readily separated from the liquid.

grease—In wastewater, a group of substances including fats, waxes, free fatty acids, calcium and magnesium soaps, mineral oils, and certain other non-fatty materials. The type of solvent and method used for extraction should be stated for quantitation.

hard detergent—A synthetic detergent which is resistant to biological attack.

ion exchange—(1) A chemical process involving reversible interchange of ions between a liquid and a solid but no radical change in structure of the solid. (2) A chemical process in which ions from two different molecules are exchanged. (3) A type of water or wastewater treatment involving the use of materials such as resin or zeolites to remove undesirable ions from a liquid and substitute acceptable ions.

iron bacteria—Bacteria that utilize iron as a source of energy or that cause its dissolution or deposition. The former obtain energy by oxidizing ferrous iron to ferric iron, which is precipitated as ferric hydrate; the latter use iron without oxidation.

Jeris rapid COD test—A chemical test to measure the relative pollutional strength of wastewater samples. Organic matter contained in wastewater is digested in an acidic oxidizing solution by heating to 165°C. This requires but a few minutes compared to the refluxing period of 1/2 to 2 hours specified by Standard Methods.

lagoon—An all-inclusive term commonly given to a water impoundment in which organic wastes are stored, stabilized, or both. Lagoons may be described by the predominant biological characteristics (aerobic, anaerobic, or facultative), by location (indoor, outdoor), by position in a series (primary, secondary, or other) and by the organic material accepted (sewage, sludge, manure, or other).

land disposal—(1) Disposal of wastewater onto land by spray or surface irrigation. (2) Disposal of solid waste materials by incorporating the solid waste into the soil by cut-and-fill techniques or by sanitary landfill operations.

leaching—(1) The removal of soluble constituents from soils or other material by percolating water. (2) The removal of salts and alkali from soils by abundant irrigation. (3) The disposal of a liquid through a non-watertight artificial conduit, or porous material by downward or lateral drainage, or both, into the surrounding permeable soil. (4) The loss of soluble constituents from fruits, vegetables, or other material into water or other liquid in which the material is immersed. (5) The escaping of free moisture from a solid waste land disposal site into the surrounding environment, frequently causing odors and other nuisance conditions of public health significance.

navigable water—Any stream, lake, arm of the sea, or other natural body of water that is actually navigable and that, by itself or by connections with other waters, is of sufficient capacity to float watercraft for the purposes of commerce trade, transportation, or even pleasure for a period long enough to be of commercial value; or any waters that have been declared navigable by the Congress of the United States.

nitrogen cycle—A graphical presentation of the conservation of matter in nature, from living animal matter through dead organic matter, various stages of decomposition, plant life and the return of living animal matter, showing changes which occur in course of the cycle. It is used to illustrate biological action and also aerobic and anaerobic acceleration of the transformation of this element by wastewater and sludge treatment.

nonionic surfactant—A general family of surfactants so called because in solution the entire molecule remains associated. Nonionic molecules orient themselves at surfaces not by an electrical charge, but through grease-solubilizing and water-soluble groups within the molecule.

nonsettleable solids—Wastewater matter that will stay in suspension for an extended period of time. Such period may be arbitrarily taken for testing purposes as 1 hour.

nutrient—A chemical substance (an element or a chemical compound) absorbed by living organisms and used in organic synthesis. The major nutrients include carbon, hydrogen, oxygen, nitrogen, sulfur, and phosphorus. Nitrogen and phosphorus are of major concern because they tend to recycle and are difficult to remove from water due to their solubility.

oils—(1) Liquid fats of animal or vegetable origin. (2) Oily or waxy mineral oils.

organic matter—Chemical substances of animal or vegetable origin or, more correctly, of basically carbon structure, comprising compounds consisting of hydrocarbons and their derivatives.

orthotolidine chlorine test—A technique for determining residual chlorine in water by using orthotolidine reagent and colorimetric standards. Under highly acid conditions, orthotolidine dye produces a yellow color proportional in intensity to the concentration of available residual chlorine. It is used for routine measurement; however, its accuracy is affected by interfering substances and color.

osmosis—The flow or diffusion through a semipermeable membrane separating unlike substances, in the course of which the concentrations of the components on the two sides of the membrane are equalized; especially the passage of solvent (usually water). In distinction from the passage of solute.

outfall—(1) The point, location, or structure where wastewater or drainage discharges from a sewer, drain, or other conduit. (2) The conduit leading to the ultimate disposal area (also see wastewater outfall).

oxidation—Chemically—the addition of oxygen, removal of hydrogen or removal of electrons from an element or compound.

oxidation process (treatment)—Any method of wastewater treatment for the oxidation of the putrescible organic matter. The usual methods are biological filtration and the activated sludge process. Living organisms in the presence of air are utilized to convert the organic matter into a more stable or mineral form.

oxygen deficiency—(1) The additional quantity of oxygen required to satisfy the oxygen requirement in a given liquid, usually expressed in milligrams per liter. (2) Lack of oxygen.

oxygen demand—(1) The quantity of oxygen utilized in the biochemical oxidation of organic matter in a specified time, at a specified temperature, and under specified conditions. (See BOD.)

oxygen depletion—Loss of dissolved oxygen from water or wastewater resulting from biochemical or chemical action.

oxygen-sag curve—A curve that represents the profile of dissolved oxygen content along the course of a stream, resulting from deoxygenation associated with biochemical oxidation of organic matter and reoxygenation through the absorption of atmospheric oxygen and through biological photosynthesis. Also called dissolved-oxygen-sag curve.

P.E.—(1) Abbreviation for population equivalent. (2) Abbreviation for Professional Engineer—a licensed individual legally able to draft plans.

population equivalent—A means of expressing the strength of organic material in wastewater. Domestic wastewater consumes, on an average, 0.17 lb. of oxygen per capita per day, as measured by the standard BOD test. This figure has been used to measure the strength of organic industrial waste in terms of an equivalent number of persons. For example, if an industry discharges 1000 pounds of BOD per day, its waste is equivalent to the domestic wastewater from 6000 persons: ($1000 \div 0.17 = 5882$ or approximately 6000).

protein—(1) Any of the complex nitrogenous compounds formed in living organisms which consist of amino acids bound together by the peptide linkage. (2) Any of a group of nitrogenous organic compounds of high molecular weight synthesized by plants and animals that, upon hydrolysis of enzymes, yield amino acids and that are required for all life processes in animal metabolism.

putrefaction—Biological decomposition of organic matter with the production of ill-smelling products associated with anaerobic conditions.

secondary wastewater treatment—The treatment of wastewater by biological methods after primary treatment by sedimentation. Common methods of treatment include trickling filtration, activated sludge processes, and oxidation.

self-purification—The natural processes occurring in a stream or other body of water that result in the reduction of bacteria, satisfaction of the BOD, stabilization of organic constituents, replacement of depleted dissolved oxygen, and the return of the stream biota to normal. Also called natural purification.

sequestering agent—A chemical that causes the coordination complex of certain phosphates with metallic ions in solution so that they may no longer be precipitated. Hexametaphosphates are an example: calcium soap precipitates are not produced from hard water treated with them. Also, any agent that prevents an ion from exhibiting its usual properties because of close combination with an added material. Also see chelating agent.

settled wastewater—Wastewater from which most of the settleable solids have been removed by sedimentation. Also called clarified wastewater.

sludge—(1) The accumulated solids separated from liquids, such as water or wastewater, during processing, or deposits on bottoms of streams or other bodies of water. (2) The precipitate resulting from chemical treatment, coagulation, or sedimentation of water or wastewater.

sludge bed—An area comprising natural or artificial layers of porous material on which digested wastewater sludge is dried by drainage and evaporation. A sludge bed may be open to the atmosphere or covered, usually with a greenhouse-type superstructure. Also called sludge drying bed.

sludge bulking—A phenomenon that occurs in activated sludge plants whereby the sludge occupies excessive volumes and will not concentrate readily.

sludge conditioning—Treatment of liquid sludge before dewatering to facilitate dewatering and enhance drainability, usually by the addition of chemicals.

sludge digestion—The process by which organic or volatile matter in sludge is gasified, liquefied, mineralized or converted into more stable organic matter through the activities of either anaerobic or aerobic organisms.

soft detergent—A synthetic detergent that responds to biological attack.

Sphaerotilus—A filamentous, sheath-forming bacterium, often considered the organism responsible for bulking sludge. In polluted streams, the presence of this bacterium is evidenced by fibrous growths adhering to rocks and plants along the stream bed.

sulfur cycle—A graphical presentation of the conservation of sulfur in nature—living animal matter through dead organic matter, various stages of decomposition, plant life, and the return of living animal matter—showing the changes that occur in this element in the course of the cycle. It is used to illustrate biological action as well as aerobic and anaerobic acceleration of the transformation of this element by wastewater and sludge treatment.

tertiary treatment—Treatment beyond normal or conventional secondary methods for the purpose of increasing water reuse potential.

thermal pollution—Impairment of water through temperature change due to geothermal, industrial or other causes.

TOC—Total Organic Carbon—A test expressing wastewater contaminant concentration in terms of the carbon content.

toxic substance—A substance that either directly poisons living things or alters their environment so that they die. Examples are cyanides found in plating and steel mill wastes, phenols from coke and chemical operations, pesticides and herbicides, and heavy metal salts. Another broad group includes oxygen-consuming substances that upset the balance of nature, such as organic matter from food plants, pulp and paper mills, chemical plants, and textile plants. Still another group consists of sulfides produced by oil refineries, smelters, and chemical plants.

turbidity—(1) A condition in water or wastewater caused by the presence of suspended matter, resulting in the scattering and absorption of light rays. (2) A measure of fine suspended matter in liquids. (3) An analytical quantity usually reported in arbitrary turbidity units determined by measurements of light diffraction.

ultimate biochemical oxygen demand—(1) Commonly, the total quantity of oxygen required to satisfy completely the first-stage biochemical oxygen demand. (2) More strictly, the quantity of oxygen required to satisfy completely both the first-stage and second-stage biochemical oxygen demands.

USPHS drinking water standards—Standards prescribed by the U.S. Public Health Service for the quality of drinking water supplied to interstate carriers and prescribed as standards by most state and local jurisdictions for all public water supplies.

wastewater outfall—Common reference to locate a wastewater discharge.

weir—(1) A diversion dam. (2) A device that has a crest and some side containment of known geometric shape, such as a V, trapezoid, or rectangle, and is used to measure flow of liquid. The liquid surface is exposed to the atmosphere. Flow is related to upstream height of water above the crest, to the position of crest with respect to downstream water surface, and to geometry of the weir opening.

zeolite process—The process of softening water by passing it through a substance known in general as a zeolite, which exchanges sodium ions for hardness constituents in the water.

zoogleal matrix—The floc formed primarily by slime-producing bacteria in the activated sludge process or in biological beds.